Meteorology

SEVENTH EDITION

RICHARD A. ANTHES
The University Corporation for Atmospheric Research
Boulder, Colorado

PRENTICE HALL
Upper Saddle River, New Jersey 07458

Library of Congress Cataloging-in-Publication Data

Anthes, Richard A.
  Meteorology/Richard A. Anthes.—7th ed.
    p. cm.—(PH earth science series)
  Includes bibliographical references and index.
  ISBN 0–13–231044–9
  1. Meteorology. I. Title. II. Series.
QC861.A58 1997   96–17574
551.5—dc20   CIP

Executive Editor: Robert A. McConnin
Production Editor: Rose Kernan
Buyer: Ben Smith
Page Compositor: Eric Hulsizer
Cover Designer: Karen Salzbach
Cover Photo: Rose Toomer

Printed in the United States of America.

Earlier editions copyright © 1992, 1985, 1980, 1976, 1971, 1967 by Merrill Publishing Company.

10 9 8 7 6 5 4 3 2 1

ISBN 0-13-231044-9

Prentice Hall International (UK) Limited, *London*
Prentice Hall of Australia Pty. Limited, *Sydney*
Prentice Hall Canada, Inc., *Toronto*
Prentice Hall Hispanoamericana, S.A., *Mexico*
Prentice Hall of India Private Limited, *New Delhi*
Prentice Hall of Japan, Inc., *Tokyo*
Simon & Schuster Asia Pte. Limited, *Singapore*
Editora Prentice Hall do Brasil, Ltda., *Rio de Janeiro*

Contents

Preface

This is the seventh edition of *Meteorology*, a book which was first written by the late Albert Miller of San Jose State College and published in 1966. Miller would hardly recognize this seventh edition because of the enormous changes in our understanding and practice of meteorology, or atmospheric sciences, in the 30 years since he wrote the first edition. In that time the world's population has nearly doubled to 5.7 billion people and the impacts of human activities on the atmosphere have become global in scale. The concentration of carbon dioxide, one of the most important greenhouse gases, has increased from 320 parts per million by volume (ppmv) to 360 ppmv. Other greenhouse gases such as methane and oxides of nitrogen have also increased significantly, and the net effect of the increase of these gases has likely been a warming of the Earth's surface. Countering the warming effect of the greenhouse gases has been the emission of aerosols, or dust, which reflect part of the sun's radiation back to space. Many scientists believe that at the present rate of building greenhouse gases, in the decades ahead the Earth will experience a climate warmer than has ever been experienced by civilization.

Human produced chemicals, chlorofluorocarbons, have also been identified as causing a significant depletion of ozone in the stratosphere. Ozone in the stratosphere protects life on Earth from the sun's ultraviolet rays. While the decreases have been relatively small in middle latitudes, the stratospheric ozone over Antarctica has decreased by more than 60% in the Southern Hemisphere winter season. This human-caused minimum or ozone hole, was confirmed by scientists in 1985.

Because of the increase of population and the amount of property at risk, the importance of weather has also increased since the first edition of *Meteorology* was published. For example, Hurricane Andrew which hit South Florida in 1992 caused more than $25 billion of property loss, the greatest loss ever caused by a hurricane in the United States. One of the worst winter storms of the century paralyzed the eastern United States in March 1993. Because of excellent forecasts by the National

Weather Service of these two extreme storms, loss of life was minimal in spite of the widespread damage. These two examples illustrate the fact that forecasting has improved significantly over the past 30 years, due largely to better observations through satellites and radars and better computer prediction models.

The seventh edition of *Meteorology* contains major revisions from the sixth edition. In addition to general updating throughout, Chapters 3–6 have been reorganized extensively with new material added, including discussions of Hurricane Andrew and the March 1993 "Storm of the Century." The seventh edition of *Meteorology* also contains new photographs, data, weather maps and figures.

I would like to thank the following people for contributing photographs or figures to *Meteorology*: Leo Ainsworth, David Baumhefner, Edward Brandes, Henry Brandli, Susan Friberg, Ginger Hein, Paul Kocin, Charles Knight, Thomas McGuire, Scott Norquay and Dennis Thomson. The National Center for Atmospheric Research* contributed a number of photographs. The photograph of Mt. Pinatubo erupting was contributed by the China Meteorological Administration. Donald Bogucki, Robert Bornstein, James Mills, Jr., David Serke and Connie Sutton provided helpful suggestions for this revised edition.

Richard A. Anthes
Boulder, Colorado

* The National Center for Atmospheric Research is sponsored by the National Science Foundation.

1
The Air Around Us

1.1 INTRODUCTION

Relative to the diameter of the earth, the atmosphere is an extremely thin gaseous envelope; if the earth were the size of a peach, the atmosphere would be thinner than the fuzz on the peach. Yet all life on earth depends on this thin veil. It regulates global temperature and prevents the occurrence of extremely high or low temperatures (such as exist on the moon, which of course has no atmosphere). In addition to the use of its constituents in biological processes, the atmosphere controls life in many ways. It acts as an umbrella or shield, filtering various types of electromagnetic radiation and high-energy particles from the sun and space. Most meteorites are burned up before they can penetrate to the earth's surface. Winds transport heat and moisture, and, in the process, mix the air and create more uniform conditions on Earth than would otherwise exist. The same winds drive the ocean currents, produce waves, erode the soil and transport pollen and insects. Weather destroys human structures and disrupts widespread systems of communication and transportation. The sounds we hear, the scents we smell, and the sights we see are all affected by the state of the atmosphere.

Meteorology is a science that seeks a more complete understanding of the physical processes that determine weather and climate. From this improved understanding comes the important practical application of prediction of future weather, ranging from short-range warnings of severe weather to long-range outlooks of seasonal temperatures and precipitation. In addition, meteorology can provide much needed information about the interactions between human activities and the environment; for example, how climate and air quality are modified by industrial and agricultural practices.

Prediction is a fundamental task of all sciences. Yet, after more than a hundred years of public weather forecasting—and in spite of steady improvements—forecasts are still the subject of countless jokes. For a period of one to three days, the accuracy

1

Figure 1.1   Photograph of a thunderstorm at night taken by David Baumhefner.

of weather forecasts is high—though certainly not perfect—but beyond a few days, the reliability falls off markedly.   Yet meteorologists deal with the same set of laws used by other physical scientists. If the astronomer can forecast an eclipse years ahead without a miss, why can't the meteorologist foretell exactly when tomorrow's rain will begin?

We hope the answer to this question will become clear in this book. The complexity of weather patterns is so great that it will never be possible to completely describe the state of the atmosphere, let alone forecast its future condition in detail. The motions of the atmosphere are composed of convective "cells" and vortices (whirlpools) of many sizes, one superimposed on another. The "chaotic" appearance of lake or ocean waves on a windy day would be more than equaled in the atmosphere if air motions could be seen. Yet each whirl plays a role in the total weather picture. It is perhaps not surprising that progress in understanding the atmosphere so that its behavior can be predicted has been painfully slow.

Existence in harmony with the environment may be natural for most forms of life, but it is not for humans with their complex social and technological systems. Climatologists have long been concerned with using knowledge about the atmosphere's characteristics to maximize agricultural production. More recently, with the explosive growth of the human population, scientists have become aware that emission of trace gases by human activities and changing land-surface characteristics such as deforestation and urbanization can significantly affect the climate and weather. These changes are discussed in Chapter 8.

This book concerns primarily the weather phenomena that occur in the lowest 10 kilometers (6 miles) of the atmosphere. After this introductory chapter, which deals with the general properties of the atmosphere and measurements, two basic concepts are employed in the discussion of atmospheric processes. One is that the atmosphere is a giant *heat engine*. An engine transforms energy from one type to another. In the atmosphere, radiant energy from the sun is transformed to heat. Because the heat energy of the atmosphere varies from place to place, some of it is changed into kinetic energy; i.e., energy of motion. Gasoline engines work the same way, of course. If the gases in the cylinder of a gasoline engine were not hotter than those on the outside, the pistons would not move. Examination of the ways in which different energy levels are created within the atmosphere is a convenient way to decipher the complex processes.

The other concept used in this book is that atmospheric processes and motions exist in a large range of sizes or **scales**.* In the case of air motion, for example, there exists a hierarchy of flow systems that range from giant "eddies," which may cover 10 percent or more of the area of the globe, to tiny whirls, which scatter the dust on a road. Although there is an interplay between each size and its smaller and bigger "brothers," they differ in their characteristics of air motion and weather and in the relative significance of the various atmospheric forces. For example, the circulation pattern of a middle-latitude cyclone has a horizontal dimension about the same as the width. In the case of the cyclone, the Earth's rotation is a significant factor in determining the flow. This is not so in the case of the thunderstorm convective cell.

1.2   PROPERTIES OF THE ATMOSPHERE

Origin and Composition

The atmosphere that exists today evolved slowly over millions of years after the formation of the Earth, which occurred approximately 4.5 billion years ago. As the Earth slowly cooled, gases that had been dissolved in the molten rock were released to form the primitive atmosphere. Volcanoes contributed to this mixture of gases, which was considerably different from the mixture of gases that comprise today's atmosphere. Table 1.1 shows the composition of gases emitted from present-day Hawaiian volcanoes and the composition of today's lower atmosphere. Because there is evidence that the composition of the earliest volcanoes was similar to that of present volcanoes, the original atmosphere must have undergone considerable transformation to reach the benign, life-supporting mixture of gases that we have today.

Most of the **water vapor** in the early volcanic eruptions condensed, filling the ocean basins. The carbon dioxide reacted with minerals to form carbonates, while much of the hydrogen escaped the Earth's gravitational field. Free oxygen probably formed after the first quarter of the Earth's life and after the formation of the first life, which

*In this book, words that are printed in bold face are important vocabulary words.

TABLE **1.1**   **Percentage by volume of gases emitted by Hawaiian volcanoes and of the present atmosphere near the surface.**

Gas	Hawaiian Volcanoes	Present Atmosphere
Water vapor (H_2O)	79.3	0–4.0 (variable)
Carbon dioxide (CO_2)	11.6	0.035
Sulfur dioxide (SO_2)	6.5	0–0.0001 (variable)
Nitrogen (N_2)	1.3	78.08
Hydrogen (H_2)	0.6	0.00005
Oxygen (O_2)	—	20.95
Argon (A)	—	0.93
Ozone (O_3O)	—	0.000007
Other	0.7	—
Total	100.0	100.0

consisted of simple anaerobic bacteria. These bacteria, which continue to exist today, lived without oxygen. Later, primitive green plants such as algae evolved. These simple organisms could produce oxygen through photosynthesis, in which carbon dioxide and water to combine to produce carbohydrates and oxygen according to the reaction.

$$6CO_2 + 6H_2O \rightarrow C_6H_{12}O_6 + 6O_2 \tag{1.1}$$

It has been estimated that 95 percent of the total oxygen was produced in this way.

In the lowest 80 kilometers of the atmosphere, the gases of the atmosphere are relatively well mixed. In this layer, known as the **homosphere,** the concentration of each constituent gas, with few exceptions, is fairly constant throughout. In contrast, in the **heterosphere,** above 80 kilometers, the various gases have tended to stratify in accordance with their weights, as occurs with liquids of different densities.

The Lower Atmosphere

In addition to the major constituents listed in Table 1.1, a host of other gases such as neon, helium, methane, krypton, xenon and oxides of nitrogen together comprise about a hundredth of 1 percent. The chemical properties of these gases are of considerable interest to the biologist because some, such as nitrogen, oxygen, and **carbon dioxide**, are involved in life processes. Many of these trace gases are also of great interest because of the way they affect the radiation budget of the Earth, and hence the climate. Carbon dioxide, water vapor, and other gases present in even smaller concentrations, such as methane and oxides of nitrogen, are called **"greenhouse gases"** because their effects on radiation received from the sun and emitted by the earth cause the surface of the Earth to be much warmer than it would be in the absence of these gases.

Because of the greenhouse effect, the Earth is 35°C warmer than it would otherwise be; without this effect Earth would be an ice-covered planet. The climates of neighboring planets Mars and Venus illustrate dramatically the importance of the greenhouse effect. With an atmosphere consisting of 97% carbon dioxide, Venus has a surface temperature of 480°C (900°F), a true "runaway greenhouse effect." In contrast, Mars with an atmosphere less than 1% of Earth's atmosphere has a minimal greenhouse effect, and temperatures even at the Martian equator vary from 10°C (50°F) at noon to –50°C (–58°F) at night. As discussed in Chapter 8, rapid increases in many of the greenhouse gases are likely to cause significant changes in the climate in the decades ahead.

Air always contains some water in the gaseous state, and sometimes the water vapor occupies as much as 4 percent of the volume. The amount, however, varies greatly in time and space. Water is the only substance that can exist in all three states—gas, liquid, and solid—at the temperatures that exist normally on the Earth. The cycle of transition between these states goes on continuously and plays an important role in maintaining life. In addition, these *atmospheric phase changes* of water play another role that is significant to the meteorologist. During the transition from a liquid or solid to a vapor state, water molecules take up some heat energy, which they obtain from their surroundings. When they revert to the liquid or solid state, they release the same amount of energy to their environment. Thus, heat consumed at one place during evaporation may be released at an entirely different place during condensation. This is an effective way of transporting heat over great distances.

Ozone is found in very minute quantities near the surface of the Earth, usually comprising less than a few parts in a hundred million. If all the ozone in the atmosphere could be brought down to sea-level pressure and temperature, it would form a layer only about 3 millimeters thick. Although the concentration of ozone is low at all levels of the atmosphere, there is a peak near the altitude of 25 kilometers where the concentration reaches 1 part in a hundred thousand. Despite the small quantities, ozone is quite significant in the radiant energy transfer that goes on in the atmosphere. Because of ozone's strong absorption of ultraviolet light from the sun, very little of this lethal radiation arrives at the surface of the earth. The ozone (O_3) of the atmosphere forms when an atom of oxygen (O), a molecule of oxygen (O_2), and a third "catalytic" particle, such as nitrogen, collide. The atomic oxygen is formed in the atmosphere by the splitting of molecular oxygen under the action of very short-wave solar radiation. The maximum of ozone near 25 kilometers is due to a balance of two factors—the availability of very short-wave solar energy to produce atomic oxygen, which is gradually depleted as it traverses the upper layers of the atmosphere, and a sufficient density of particles to bring about the collisions required.

As discussed in Chapter 8, in recent years major losses of **ozone** have occurred over Antarctica in the Southern Hemisphere springtime. These losses have been caused by chemical reactions involving human-made chlorofluorocarbons—CFCs—through complex reactions involving minute quantities of ice in the polar stratosphere.

Fairly high concentrations of ozone sometimes occur in the lowest few hundred meters of the atmosphere, especially over urban areas. Ozone, a corrosive toxic gas, is an important constituent of the so-called "photochemical smog" that afflicts some large cities. The atomic oxygen required for the reaction described above is formed in smog principally through the action of solar radiation on nitrogen dioxide, a product of combustion.

A variety of solid particles are suspended in the air. These include fine dust particles swept up by the wind from exposed soils; soot from forest fires, industrial fires, industrial plants, and volcanoes; pollen and microorganisms lifted by the wind; meteoritic dust; and salts injected into the atmosphere when ocean spray is evaporated. Large particles are too heavy to remain long in the air, but there are many, so many that they cannot be seen individually with the naked eye, that remain suspended for months or even years. The minute particles of dust thrown high into the atmosphere by the violent eruption of the volcano Krakatoa in the East Indies in 1883 circled the globe for at least two years, producing magnificent sunrises and sunsets. The eruption of El Chichón in Mexico on 28 March 1982 injected massive amounts of dust and sulfur dioxide into the atmosphere (Figure 7.10). As this dust circled the Earth, solar radiation was reduced by as much as 20 percent at the surface in some locations, perhaps causing a decrease of surface temperature of about 1°C in the months following the eruption. And more recently, the eruption of Mt. Pinatubo in the Philippines on June 15, 1991 (Figure 1.2), one of the largest this century, injected 20-30 Teragrams of sulfur dioxide (SO_2) into the lower stratosphere at altitudes of up to 30 km. This SO_2 was oxidized to sulfuric acid and condensed as stratospheric aerosols. Although initially confined to the tropics, within one year the aerosols had spread to the entire globe, and aerosol concentrations of one to two orders of magnitude above back-

Figure 1.2   Photograph of an eruption of Mt. Pinatubo on June 15, 1991.

ground values persisted for several years. These aerosols affected the atmospheric chemistry of the stratosphere, causing a significant (5–10%) depletion of ozone. In addition, through their reflection of sunlight, they produced a global cooling of about 0.5°C, an effect which lasted several years.

   Many of the small dust particles in the air act as centers, or **nuclei**, around which minute water drops or ice crystals form. (This will be discussed a little later on in Chapter 2, *Clouds and Precipitation*.) Figure 1.3 shows the sizes of these nuclei and, for purposes of comparison, the sizes of air molecules and liquid and solid water particles.

   Dust particles in the air, as well as water droplets and ice crystals, affect the transparency of the air. They not only reduce visibility, but also they prevent some of the sun's energy from penetrating to the surface of the earth. There has been some speculation that humans, who have been increasing the atmosphere's load of dust with factories and high-flying aircraft, may be changing the climate. Although humans are undoubtedly the biggest contributors of the grime in urban areas (where the number of dust particles can reach as high as several million per cubic centimeter in smoke-laden air), our contribution to the overall dustiness of the atmosphere is small (at present) compared to such natural sources as volcanoes.

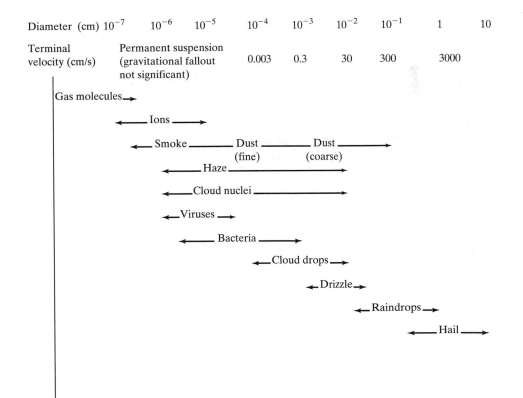

Figure 1.3   Diameter and terminal velocities (maximum fall velocities) of particles in the atmosphere.

The Upper Atmosphere

The major constituents of the atmosphere remain virtually unchanged up to 80 or 90 kilometers, although there are significant variations in such minor constituents as ozone, dust, and water vapor. But above this level (at which point only 0.0002 percent of the total atmosphere remains), the relative amounts and types of gases change. The gases of the "thin" air of the heterosphere undergo various "photo-chemical" effects induced by the very short waves of ultraviolet and X-ray radiation from the sun. As a result of these photochemical reactions, molecular oxygen is split into two atoms and many molecules and atoms are ionized. That is, electrons have been ejected from their atoms, leaving them with an overall positive charge.

The entire layer from about 80 kilometers upward contains a large number of positively charged ions and free electrons and is therefore referred to as the **ionosphere**. This electrically charged portion of the atmosphere is very useful for radio communications, since it reflects radio waves. Around-the-world transmissions are accomplished by bouncing radio waves, which move in straight lines, between the ionosphere and the Earth's surface.

The distribution of electron density with height fluctuates a great deal. There is a fairly regular daytime-to-nighttime change in the strength of some layers due to the changes in intensity of the solar radiation. In addition, there are occasional sudden ionospheric disturbances (S.I.D.s) and "ionospheric storms" that are associated with disturbances on the sun. An S.I.D. lasts from 15 to 30 minutes, and it is produced by bursts of ultraviolet energy from the sun, causing a sudden increase in the production of electrons. Because electrons absorb part of the radio energy that strikes them, a sudden increase in their number may actually smother the radio energy, leading to fadeouts of communications on the sunlit side of the Earth. Ionospheric storms, which can occur during the day or night and last for hours or even days, are believed to be caused by a stream of charged particles emitted from the sun. These fast-moving particles, guided toward the poles by the Earth's magnetic field, not only ionize the air, but also they produce the beautiful displays of aurora borealis (northern lights) and aurora australis (southern lights).

Temperature Distribution in the Vertical

The mean temperature distribution in the vertical shown in Figure 1.4 provides a basis for dividing the atmosphere into shells or layers. In the lowest of these layers, the **troposphere**, the temperature decreases with height, on the average, at the rate of 6.5°C/km (3.5°F/1000 ft). In this layer, vertical convection currents, induced primarily by the uneven heating of the layer by the Earth's surface, keep the air fairly well stirred. Practically all clouds and weather and most of the dust and water vapor of the atmosphere are found in this turbulent layer. Its upper boundary, called the **tropopause**, is at an average elevation of about 10 kilometers, but it varies with time of year and latitude and even from day to day at the same place. Typically, the tropopause is at an elevation of 15 or 16 kilometers over the equator and only 5 or 6 kilometers over the polar regions. It tends to be higher in summer than in winter.

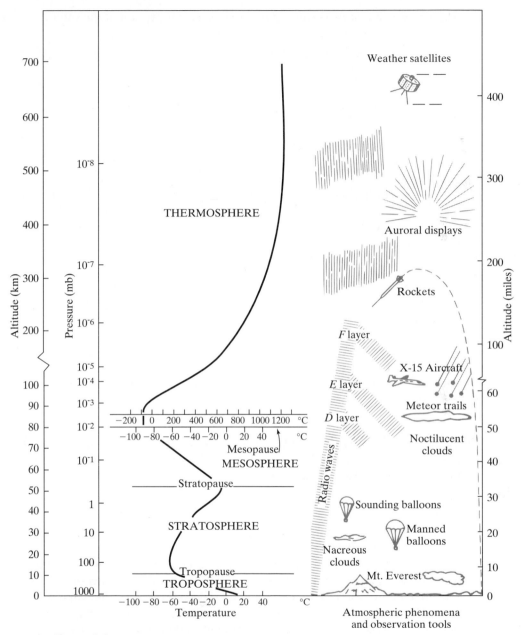

Figure 1.4   Vertical distribution of atmospheric temperature and phenomena.

In the **stratosphere**, with an upper boundary at about 50 kilometers, the temperature is relatively constant in the lower part; in the upper part it increases with height, reaching a temperature at the **stratopause** that is not much lower than at sea level. The warmth of this layer is due to the direct absorption of the sun's ultraviolet rays by ozone. It will be shown later that this increase of temperature with height inhibits the air from moving up and down. As a result, this layer acts as a lid on the turbulence and vertical motions of the troposphere. Clouds and vertical convection currents formed near the Earth's surface do not usually penetrate very far into the stratosphere. The air in this layer is also quite dry, since it is largely cut off from the sources of moisture at the Earth's surface.

The **mesosphere** is the zone between 50 and 85 kilometers in which the temperature decreases rapidly with height, reaching about –95°C at the **mesopause**, which is the coldest point in the atmosphere. Significant vertical movement of air exists in the mesosphere.

Above the mesosphere is the **thermosphere**, a layer in which the temperature increases rapidly, then more slowly with height. This increase in temperature is due to the absorption of short-wave solar radiation by atoms of oxygen and nitrogen. However, the high temperatures of the thermosphere can be misleading when compared to similar temperatures near the Earth's surface. Temperature is related to the average speed of the molecules; thus, in the thermosphere the molecules are moving very rapidly. However, the density is so low that very few of these fast-moving molecules would collide with an object located at this level, hence very little energy (heat) would be transferred. Thus, paradoxically, the thermosphere is both very hot and very cold.

At heights above 500-600 kilometers, the density of particles is so low that collisions among them are infrequent, and some of the particles can escape the gravitational pull of the Earth. This zone, which marks a transition from the Earth's atmosphere to the very thin interplanetary gas beyond, is called the **exosphere**.

1.3 ATMOSPHERIC MEASUREMENTS

Characteristics of the Gases

There are significant variations in space and time of almost all the physical properties of the atmosphere. However, some characteristics, such as air temperature, water content, and air motion, account for such a large proportion of the total energy and essentially all of everyday "weather" that they deserve special consideration in this chapter. Because the atmosphere is composed mostly of gases, we shall review briefly the behavior of gases and the measurements that are used to describe their physical state.

First, matter can exist in the form of a solid, a liquid, or a gas. Unlike solids and liquids, gases can expand or be compressed easily, a property very important to some atmospheric processes such as thunderstorm formation. One important property of a gas is the space that a given number of gas particles may occupy. At sea level there are about 25×10^{18} molecules of air in each cubic centimeter (about the volume of the tip of your small finger up the base of the nail). The total mass of molecules per unit volume is the **density**. Near sea level, 1 cm^3 of air contains a mass of about 1.2×10^{-3} g, so the density is 1.2×10^{-3} g/cm^3.

Another bulk property of gases is pressure, which is defined as the force per unit area exerted on any surface being bombarded by the moving molecules. Held to the

Earth by gravitational attraction, the Earth's atmosphere has a cumulative force or weight per unit area averaging 14.7 lb/in.^{2} at mean sea level. In the SI system of units* the standard sea-level pressure is 101.3 kilopascals (kPa), where 1 Pa is 1 kg/m/s^2. For many meteorological purposes, including weather reports, the millibar (mb), which is 0.1 kPa, is the commonly used unit of pressure. Thus the average sea-level pressure is 1013 mb.

The fact that gases are easily compressed is illustrated by the pressure distribution with height given in Figure 1.5 and in the table of Appendix 2. In water, the pressure increases almost exactly in proportion with depth below the water surface, but not so in the atmosphere. As can be seen from the table, one must ascend 2500 meters for the pressure to fall off 25 percent (about 250 millibars) from its sea-level value, but over 3000 meters more for another 25-percent drop, and another 4800 meters for an additional 25 percent. In the layer between sea level and 5000 meters, the pressure decreases 470 millibars, but in the layer between 30,000 meters and 35,000 meters, the decrease is less than 7 millibars. Evidently, the air near the bottom of the atmosphere is compressed by the weight of the air resting above. The compressibili-

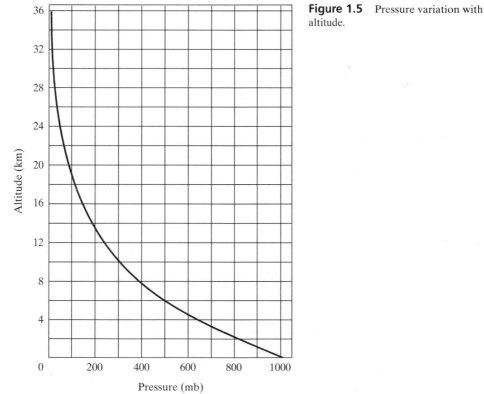

Figure 1.5 Pressure variation with altitude.

*Système. International d'Unités (designated SI in all languages).

ty characteristic of gases is of great significance in atmospheric processes. As will be pointed out in a later chapter, rapid compressions and expansions of air occur naturally in the atmosphere due to vertical displacements, and these are largely responsible for much of the weather.

Gas Laws

Two experimental laws relate the properties of temperature, density (or volume), and pressure in gases. Boyle's law (named after Englishman Robert Boyle) states that if the temperature of a gas does not change, its pressure varies directly as the density varies,

$$p = c_1 \rho \text{ (at constant temperature)} \qquad (1.2)$$

where p is pressure, ρ is density and c_1 is a constant. According to Boyle's law, a container of gas containing twice as many molecules as another container of the same volume will have twice the pressure, assuming that the temperatures of the two gases are the same.

The relationship between temperature of a gas and its volume, when the pressure is the same, is given by Charles law (named after the French scientist Jacques Charles),

$$V = c_2 T \text{ (at constant pressure)} \qquad (1.3)$$

where c_2 is another constant. Thus, according to Charles' law, a gas at two times the temperature of another gas will occupy twice the volume if the pressure of the two gases is the same.

Boyle's law and Charles' law may be combined to relate pressure (p), temperature (T), and density (ρ) through the **equation of state**,

$$p = \rho RT \qquad (1.4)$$

where R is the gas constant for dry air. The value of R is 287 joules per kilogram per Kelvin (J/kg/K).* As indicated by the equation of state, at the same pressure, cold air is denser than warm air.

Observations of the Atmosphere

Weather observations have been made by humans since earliest times, but systematic measurements of the elements did not begin until the invention of instruments during the seventeenth and eighteenth centuries. Until the twentieth century, measurements were confined to the air close to the ground. Systematic measurements of most of the Earth's atmosphere are scanty even today.

Measurement of the state of the atmosphere is quite difficult. In addition to the usual requirement that an instrument measure accurately whatever it is designed to measure, the meteorological instrument must be rugged enough to withstand the force of buffeting winds, the corrosive action of high humidity and flying dust, the extremes of heat and cold. Another difficulty is the inaccessibility of much of the atmosphere, so that instruments must be built to transmit their measurements to dis-

---

*A joule is a unit of energy and is equal to 4.187 calories.

tant ground points: They must be rugged and light enough to be carried aloft by balloons, rockets, and satellites and cheap enough so they can be used in the large quantities necessary to observe the atmosphere. Finally, meteorological measurements must be "representative," a difficult objective to achieve, considering the enormous size of the atmosphere and the comparatively few observations that can be made. For example, consider the difficulty in estimating the correct distribution of rainfall over a region. Taking the depth of water in a single 8-inch-diameter rain gauge as representative of the average rainfall over an area of many square miles is somewhat like assuming that the height of a single random person of a certain age is equal to the average for all people of the same age.

Temperature

The most common device for measuring the temperature of the air is the liquid-in-glass thermometer. As the air temperature increases and the heat is transferred to the thermometer, the liquid expands; as the air and the thermometer cool, the liquid contracts. Liquid-in-glass thermometers that will register the maximum or minimum temperature during a period require slight modifications. The maximum thermometer has a constriction of the bore of the glass tube just above the bulb; as the temperature rises, the mercury is forced through the constriction, but when the temperature falls, the weight of the mercury in the column is insufficient to reunite it with that in the bore. Thus, the top of the mercury column indicates the highest point reached. The maximum thermometer can be reset by shaking the thermometer, thereby forcing the mercury through the constriction.

The minimum thermometer contains alcohol in the bore, with a small, dumbbell-shaped glass index placed inside the column of alcohol. The index is kept just below the meniscus of the alcohol column by surface tension. With the thermometer mounted horizontally, when the alcohol contracts, the meniscus drags the index with it; but when the alcohol expands, the meniscus advances, leaving the index at its lowest point. To reset, the index can be returned to the meniscus merely by tilting the thermometer.

Expansion-type thermometers are also used for recording the temperature. A bimetal thermometer can be used to move a pen arm that traces its position on a paper chart driven by a clock. The bimetal thermometer is the type ordinarily used in thermostatic control devices. Two strips of metal, having different rates of expansion during a temperature change, are welded together and rolled. The difference in expansion of the two strips causes changes in the curvature of the elements as the temperature changes; with one end fixed in position, the other end is free to move the pen arm and indicate the temperature.

The principal use of electrical thermometers in meteorology is in the **radiosonde**, which is attached to a balloon and transmits temperature, pressure, and humidity data by radio as it ascends through the atmosphere. There are two general types of electrical thermometers: (1) the thermoelectric thermometer, which operates on the principle that temperature differences at the junctions of two or more different metal wires in a circuit will induce a flow of electricity, and (2) the resistance thermometer, which is based on the principle that the resistance to the flow of electricity in a substance depends on its temperature. It is the latter that is used in the radiosonde. The ceramic elements commonly used are called thermistors.

Obtaining meaningful air temperatures is not a simple procedure. Air is a poor conductor of heat and quite transparent to radiation, especially in the short wavelengths emitted by the sun. That this is so is quite evident when one moves a few feet from the shade into the sun; even though the air temperature is almost identical, one feels much warmer in the sun.

Most thermometers are much better absorbers of radiation than air. They absorb energy that passes right through the air from the sun and other warm objects. If the thermometer is to measure the air temperature, such radiation must be prevented from reaching the thermometer. This is accomplished by shielding the thermometer, at the same time keeping it in contact with the air. This can be done by enclosing the thermometer within a highly polished tube, allowing plenty of room for air to circulate past the thermometer. However, when several temperature-measuring devices are used, it is convenient to house them in a special "instrument shelter," which permits air to pass through. The shelter also keeps the instruments dry during rain. This is important because a wet thermometer will generally read lower than a dry one. (See the section on humidity.)

The thermometer should be ventilated artificially when there is little wind because the conductivity of air is poor ("dead," or stagnant, air is often used for insulation), and the thin layer that encases the thermometer might have a different temperature than the "free" air. By stirring the air, this layer is mixed with the surrounding air.

Even if the precautions given above are taken in measuring temperature, there is still the question of how to interpret temperature measurements. On a sunny windless day the temperature of air within an inch or two of a cement sidewalk can be 30°F or more higher than at the 4-foot level. Even at the same height above the ground, the temperature differs greatly between the city and the country, within forests and over open land, along sloping land and over flat land. Variations in temperature over small distances are so important that measurements at a single point can rarely be considered representative of the average conditions closer than 2 Fahrenheit degrees.

Pressure

Pressure in a fluid is defined as the force per unit area exerted on any flat surface. The orientation of the surface will not affect the pressure. In the case of the atmosphere, which has no outer walls to confine its volume, the pressure exerted at any level is due almost entirely to the weight of the air pressing down from above; i.e., the force results from gravitational attraction. (The units of pressure were given in the section on characteristics of gases.)

As shown in Figure 1.5, the pressure changes rapidly in the vertical. In the lowest few kilometers, the pressure decrease amounts to about 1 millibar per 10 meters. Because of compressibility of air, the rate at which the pressure decreases with height becomes slower at greater heights.

Variations of pressure in the horizontal are much smaller than they are in the vertical. Near sea level, the change of pressure with distance rarely exceeds 3 millibars per 100 kilometers and is usually much less than this rate. Although the horizontal variations in pressure are small, they are responsible for the winds we observe. Since pressure measures the weight per unit area of the atmosphere, variations in pressure along any horizontal surface (such as sea level) must arise through variations in the average density of the atmosphere; i.e., there must be more molecules in a column of air above a point having high pressure than in one above a point where low pressure is observed.

Pressure also changes with time at a single place. In **extratropical** regions, the greatest part of these changes is of an irregular nature, caused by occasional invasions of air having a different mean density. But there is also a smaller contribution to the pressure change, a quite regular **diurnal** oscillation that contributes two peaks (at about 10 A.M. and 10 P.M.) and two minima (at about 4 A.M. and 4 P.M.). The difference between maxima and minima is greatest near the equator (up to 3 millibars), decreasing to practically zero in the polar regions. These regular fluctuations in the pressure are analogous to the tidal motions in the ocean, but in the case of the ocean it is the gravitational pull of the moon and sun that causes the water surface to bulge slightly outward from the Earth, while in the atmosphere the daily heating and cooling cycle is the dominant cause of the pressure variations. Diurnal wind oscillations accompany the migration of these maxima and minima of pressure around the Earth each day; these are hardly detectable at low elevations in the atmosphere, but they become quite strong between 80 and 100 kilometers.

The mercurial barometer, invented by Torricelli in 1643, is still the fundamental instrument for measuring atmosphere pressure. It is constructed by filling with mercury (Hg) a long glass tube sealed at one end, inverting the tube, and placing the open end into a dish of mercury. The mercury in the tube will flow into the dish until the column of mercury is about 30 inches high (at sea-level sites), leaving a vacuum at the top. In principle, the barometer is merely a weighing balance (Figure 1.6), the pressure exerted by the atmosphere on the exposed surface of the mercury in the

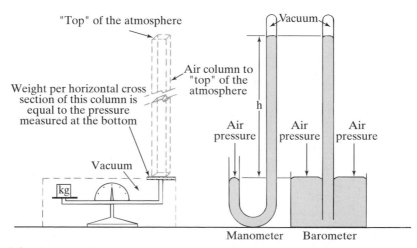

Figure 1.6   Principle of barometer.

dish equaling that exerted by the mercury in the tube. Changes in atmospheric pressure are detected from changes in the height of the column of mercury. Although it is now the custom to use the height of the column as a pressure unit (millimeters or inches of mercury), conversion to such units as millibars can be made as follows.

The density of mercury at 0°C is 13.6×10^3 kg/m³. Note that the height of the column of mercury will depend on temperature as well as pressure, since mercury expands with increased temperature. To obtain the true pressure, one must correct for this expansion. The mass of a column of mercury = mercury density × volume = density × height × cross-sectional area of tube. Its *weight*, therefore, would be obtained by multiplying by the acceleration of gravity (weight = mass × gravity) and the weight per unit area (pressure) exerted by the column would be obtained by dividing by the area. Thus, pressure = gravity × density × height. For example, if the height of the column of mercury were 76 centimeters (0.76 m), the pressure p would be

$$
\begin{aligned}
p &= (9.8 \text{ m/s}^2) \times (13.6 \times 10^3 \text{ kg/m}^3) \times (0.76 \text{ m}) \\
&= 101.29 \times 10^3 \text{ kg/m/s}^2 \\
&= 101.29 \text{ kPa} \\
&= 1012.9 \text{ mb}
\end{aligned}
\tag{1.5}
$$

Here a value of 9.8 m/s² has been used for gravity. In practice, since gravity varies slightly, the value at the particular place should be used.

The aneroid barometer, although not usually as accurate as the mercurial barometer, is more widely used because it is smaller, more portable, usually cheaper to manufacture, and simpler to adapt to recording mechanisms. Its principle of operation is that of the spring balance (Figure 1.7). A thin metal chamber, with most of its air evacuated, is prevented from collapsing under the force of atmospheric pressure by

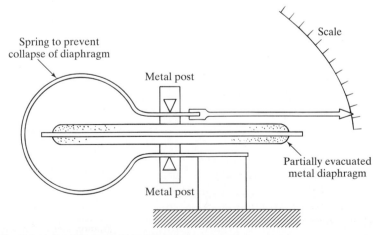

Figure 1.7 Principle of the aneroid barometer.

a spring. The force exerted by the spring depends on the distance it is stretched. The balance between the spring force and the atmospheric force will thus depend on the width of the chamber. Changes in this width can be discerned by movement of an arm attached to one end of the chamber; these deflections are usually magnified by levers. If a pen is attached to the arm, the instrument becomes a barograph.

The altimeters used in aircraft and by mountain climbers, surveyors, and others are usually nothing more than aneroid barometers made to indicate altitude rather than pressure. They are designed to give the altitude for the standard ("normal") pressure distribution with height, and so will give slightly erroneous readings. For accurate determinations of altitude, the true density of the air for each altitude increment must be measured and corrections to the indicated altitude must be computed.

Humidity

The concentration of water vapor in the atmosphere varies from practically zero to as much as 4 percent (4 grams of water in every 100 grams of air). The extreme variability in the amount of water vapor in both space and time is due to water's unique ability to exist in all three states—gas, liquid, and solid—at the temperatures normally found on Earth. Water vapor is continuously extracted from the atmosphere through condensation (vapor to liquid) and sublimation (vapor to ice); some of this may fall to the Earth through precipitation. Water is continuously being added to the atmosphere through evaporation (liquid to vapor) from oceans, lakes, rivers, soil, plants, and raindrops, and also through sublimation (ice to vapor) from snow and ice.

The exact amount of water vapor that exists at any place and time is important to the meteorologist because of the role water plays in weather processes. First, it is significant because condensation is an important aspect of weather. Second, water vapor is the most important absorber of radiation in the air and thus affects the energy balance of the atmosphere (Chapter 3). Third, the release of heat during condensation is an important source of energy for atmospheric circulations.

For a substance such as water to change its phase from solid to liquid or liquid to gas, the forces that bind the molecules together must be broken down. Work must be done in overcoming these intermolecular forces, so energy must be applied from the environment. Consider a glass of water mixed with crushed ice resting in a room on a warm summer day. Stir the water-ice mixture slowly with a thermometer. The temperature of the water-ice mixture will remain at 0°C (32°F) until all the ice is melted, at which time the temperature of the water will start to increase. The energy (heat) that is being supplied by the warm environmental air is used to break down the crystalline structure of the ice (melt the ice) rather than increase the temperature of the water-ice mixture. Because this heat energy does not produce a temperature change, it is referred to as **latent** (hidden) heat.

Latent heat is required to evaporate water, melt ice or **sublimate** ice (convert ice directly to vapor). As shown in Figure 1.8, it requires approximately 600 calories of heat to evaporate one gram of water (this is why evaporation of water from your skin produces a noticeable cooling effect; the body loses approximately 600 calories of heat when it evaporates one gram of water from the skin). (Note: The calories here

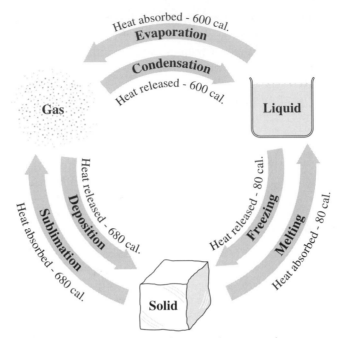

Figure 1.8   Heat exchange when water changes from one phase to another.

are gram-calories rather than the so-called "calories" generally referred to with reference to caloric content of food or the "calories" burned when exercising. The latter are kilogram-calories and are one thousand times greater than gram-calories. Thus a soft drink which is labeled as containing 100 "calories" actually contains 100 kilogram-calories or 100,000 gram-calories.)

The latent heat supplied when water is evaporated and ice melted or sublimated is given off, or released in the reversed processes of water vapor condensing into liquid water, liquid water freezing, or water vapor being converted directly to ice (**deposition**). The approximate number of calories released in these processes is given in Figure 1.8.

The quantity of water vapor in the air can be expressed in a variety of ways. One is the density of water vapor, usually referred to as the **absolute humidity** and expressed as the mass of water vapor in a given volume. Normally, there are not more than about 12 g/m^3, although as much as 40 g/m^3 can occur at high temperatures.

The *partial pressure* of water vapor (i.e., the contribution made by water to the total atmospheric pressure) is another measure that can be used. It is usually expressed in millibars or inches of mercury. Typically, the water-vapor pressure does not exceed 15 millibars, although it can reach double this value or more at high temperatures.

The amount of water vapor that can be added to a volume at any given pressure and temperature is limited. When a volume has reached its capacity for water vapor, it is said to be saturated and the volume will normally accept no more gaseous water. The **saturation vapor pressure**, as this vapor pressure is called, is a function of temperature. (See Figure 1.9.) This dependence of the saturation vapor pressure on

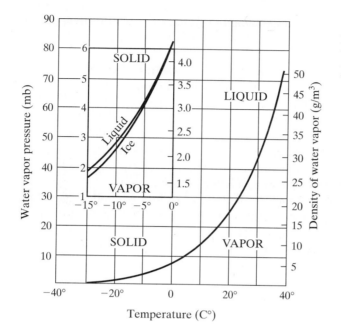

Figure 1.9 Saturation vapor pressure and density as a function of temperature. (Insert shows variation of vapor pressure and density over liquid and ice below 0°C.)

the temperature is why cooling is so important in producing condensation. For example, a sample of air having a temperature of 20°C and a water vapor pressure of 20 mb would be unsaturated, because the saturation vapor pressure at 20°C is 25 mb. If this air were cooled, the sample would become saturated at a temperature of 16°C and further cooling would result in condensation of the excess moisture. When a 4°C temperature was reached, the saturation vapor pressure would be only 10 mb; so half of the vapor would have liquefied. The temperature to which a sample of air must be cooled (at constant atmospheric pressure) to make it "saturated" is called the **dew point**. In the above example, the dew point of the sample before condensation began was 16°C; after condensation begins, the temperature and the dew point are the same. Thus, the dew point is a measure of the water-vapor pressure; the difference between the temperature and the dew point is a measure of the degree of saturation of the air.

The **relative humidity** is the ratio, expressed in percent, of the actual vapor pressure to saturation vapor pressure; i.e., relative humidity = actual vapor pressure/saturation vapor pressure multiplied by 100. Relative humidity measures how close the air is to saturation (100 percent indicates saturation). In the example given above, the relative humidity before cooling was $^{20}/_{25} \times 100 = 80$ percent; between 16°C and 4°C, the relative humidity remained at 100 percent. The relative humidity is very sensitive to temperature change. There is normally a large diurnal change of relative humidity, even when the quantity of moisture in the air is constant, merely because the saturation vapor pressure continuously changes with the daily temperature variation.

The most accurate way to measure humidity is to extract all the water vapor from an air sample (perhaps by passing it through a chemical drying agent) and then weigh the water collected. However, for meteorological observations, this procedure is not practical because the sampling time is too long and the analytical tools are too complicated. Study of the atmosphere requires almost instantaneous sampling under field conditions.

None of the many techniques used to measure atmospheric humidity is completely satisfactory; here we will mention just a few of the most commonly used instruments. The hair hygrometer is probably the oldest instrument for measuring moisture. Many organic materials such as wood, skin, and hair absorb moisture when the humidity is high, and so they expand. Blond human head hair increases its length by about $2\frac{1}{2}$ percent as the relative humidity increases from 0 to 100 percent. The hair hygrometer consists merely on one or more hairs whose changes in length are made to move a pointer or, in the case of a hygrograph, a pen.

The psychrometer consists of a pair of ordinary liquid-in-glass thermometers, one of which has a piece of tight-fitting muslin cloth wrapped around its bulb. The cloth-covered bulb, called the **wet bulb**, is wetted with pure water, and both thermometers are then ventilated. The dry bulb will indicate the air temperature, while the wet bulb will be cooled below the dry bulb temperature by evaporation when the relative humidity is less than 100 percent. The amount of evaporation, and therefore of cooling, will depend on how nearly saturated the air is. If the surroundings are saturated, there will be no evaporation, and the west and dry bulbs will read the same. The difference between the dry and wet bulbs, called the *depression of the wet bulb*, is a measure of the degree of saturation of the air. Tables can be used to obtain the relative humidity or dew point from psychrometer readings.

An electrical hygrometer is used in the radiosonde. It consists of an electrical conductor that is coated with lithium chloride, which is hygroscopic. The amount of moisture absorbed by the conductor depends on the relative humidity of the air, and the electrical resistance of the conductor is a function of its dampness.

Wind

Air in motion, or *wind*, continuously stirs the atmosphere. By transporting heat, moisture, pollutants, etc., from one place to another, it acts to redistribute the concentration of these quantities. Chapter 4 will discuss what causes winds; at this point, only some of the general characteristics of wind will concern us.

Although air moves up and down as well as horizontally, the speed of vertical displacements is usually a tenth or less of that of the horizontal component. Even though it is quite small, the vertical component is very important. As we shall see later, it is the up-and-down motion of air that is principally responsible for the formation and dissipation of clouds in the atmosphere. Only the *horizontal* component of the wind is measured on a regular basis; the much smaller vertical component must be computed from relationships between it and the changes of the horizontal wind in space.

Many instruments are used to measure the horizontal wind velocity near the surface of the Earth. The wind vane is a very old device for indicating wind direction. Because if points *into* the wind, it is customary to designate wind direction as that *from* which the air comes. Thus, when the air is moving *from* northwest (315°) to southeast (135°), the wind direction is said to be northwest (315°).

A large variety of *anemometers* exist for measurement of wind speed. The cup anemometer is probably the most widely used. It consists of three or more hemispherical cups clustered around a vertical shaft. Air striking concave sides of the cups exerts more force than that hitting the convex sides, causing the cups and therefore the shaft to turn. The number of rotations per unit time is a measure of the wind speed.

Air flow is retarded by friction with the ground and deflected by obstacles, so that the position of wind-measuring instruments must be carefully considered. At an airport, for example, both the wind speed and direction atop the control tower may be considerably different from those at the end of the runway. Typically, the wind speed increases rapidly with height near the surface, so that the height of an anemometer will greatly influence the speed recorded. Unfortunately, there is no uniformity of height for anemometers, although arbitrary standards have been set.

Gustiness and the Diurnal Wind Variation. Anyone who has watched a wind vane oscillate and a cup anemometer alternately increase and decrease its rotation speed during brief periods of time, or who has watched a flag flutter in the wind, can attest to the normal unsteadiness of the wind. An example of such fluctuation can be seen from the recording of the wind direction and speed shown in Figure 1.10. These velocity changes are attributed to the fact that air normally does not move in straight lines, but in irregular paths. We call such erratic airflow patterns turbulent. Successive particles passing a single point in space may have had distinctly different histories, some having been most recently at higher elevations than the point, others at lower elevations. Normally, those coming from higher up will arrive with relatively low speeds. The result will be gusty winds at the observation point.

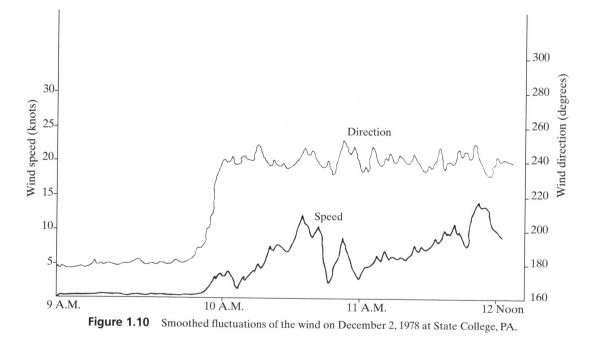

Figure 1.10   Smoothed fluctuations of the wind on December 2, 1978 at State College, PA.

On clear nights strong radiative cooling at the ground often produces there a layer of very cold air which is overlain by warmer air aloft (a temperature **inversion**). Under such conditions the air near the surface is very stable, vertical mixing is inhibited, and the surface wind is often nearly calm or shows a light flow from a direction determined by local effects such as sloping terrain. The light southerly wind between 9 and 10 A.M. in Figure 1.10 is an example of the winds under such stable conditions. Shortly before 10 A.M. on this day, surface heating destroyed the inversion, and air from higher aloft was mixed down to the surface. This larger scale current of air was predominantly from the west, and so the surface wind direction shifted when the inversion was eliminated. As the vertical stability is reduced by surface heating, the winds become more turbulent. Therefore, strong gusty winds are most likely to occur near midday when the solar heating is at a maximum.

Upper-Air Observations

Sounding Techniques.   Winds at levels above the reach of ground-based instruments are measured by tracking helium-filled or hydrogen-filled balloons. The horizontal displacements over short intervals of time as the balloon ascends give the velocity. The changes in position of the balloon may be determined by any of these methods: (1) optically, by the use of a theodolite (similar to a surveyor's transit); (2) by reflection of radio waves (radar) from a target carried by the balloon; (3) by tracking of the radio signal transmitted by a radiosonde carried by the balloon; and (4) by tracking the balloon using GPS (Global Positioning System) satellites.

Systematic measurements of meteorological conditions in the free atmosphere high above the surface began around the turn of the century. Until 1938, when the radiosonde came into use, the sounding instruments had to be retrieved before the data became available. Instruments recording temperature, humidity, and pressure were carried aloft by balloons, kites, and airplanes. In the case of a free balloon, the instrument dropped to the Earth on a parachute, and the processing of the data had to wait until some finder returned the instrument. Kites were extremely laborious to handle and they rarely reached heights greater than 3 kilometers. Airplane soundings were expensive and, at least in the early days, could not provide data to the altitudes desired or during periods of severe weather.

The radiosonde, carried aloft by balloons, transmits its measurements by radio back to a ground station. The radiosonde used in the United States consists of a lightweight, inexpensive radio transmitter that emits a continuous signal. The temperature-measuring and humidity-measuring elements control the frequency or amplitude (intensity) of the audio output of the radio signal. An aneroid barometer cell, moving a contact arm across a series of metal strips, alternately connects temperature and humidity into the circuit. By setting consecutive contacts for low pressure intervals, the temperature and humidity are recorded as functions of pressure. The altitude can be computed if the vertical distribution of temperature, humidity, and pressure is known. The radiosonde is the principal tool of the meteorologist for systematically observing conditions of the lowest 30 kilometers of the atmosphere.

However, recent techniques have been developed to measure the vertical distributions of water vapor and temperature from the ground. These techniques include use of instruments that measure the different wavelengths of radiation emitted by the atmospheric layers above. This radiation depends on the temperature and humidity structure, and thus is an indirect measure of these variables. Wind profiles may be obtained by remote-sensing devices located on the ground. These devices are radars which measure the velocities of turbulent air motions that result from movement of the wind at the different levels. One of the great advantages of these remote-sensing systems is that they provide essentially instantaneous vertical profiles which are continuous in time, rather than data obtained at widely spaced intervals (12 hours for conventional radiosonde observations). The continuous observations provide for a much more complete description of the atmosphere.

Satellites

Observations of most of the Earth's atmosphere are inadequate. About 70 percent of the Earth's surface is covered by oceans, and a large proportion of the rest is dominated by mountains, snow, deserts, and jungles. Even in populated areas, the density of weather-observing stations permits the construction of only a very "coarse-grained" picture of the atmosphere.

Weather satellites have improved this situation considerably. Equipped with cameras, they transmit to Earth pictures of the clouds as seen from above. These photographs have been valuable, especially over the oceans, to pinpoint the locations of storms. On a few occasions, hurricanes (Figure 5.12) that were undetected by the low-density oceanic network of stations have been uncovered by weather satellites. Measurements of radiation from the Earth have also increased knowledge of the Earth's energy balance (Chapter 3).

There are both polar-orbiting satellites, which view strips of the Earth as they complete each orbit, and geostationary satellites, which remain over a fixed spot on the equator. This is possible when the satellite is so placed that it makes one revolution around the Earth in the same time it takes the Earth to make one rotation. An advantage of the geostationary satellite is that the evolution of weather systems with time can be seen by comparing pictures of the same area at short time intervals (usually $\frac{1}{2}$ hour). Time-lapse movies of these pictures reveal the spinning cyclones, the growth of thunderstorms, and the translation of all weather systems across the globe. Figure 3.6 shows a photograph from the geostationary satellite.

While the visible satellite photographs reveal much qualitative information about weather systems, satellites are also providing much-needed quantitative data, especially over remote ocean areas where conventional observations are scarce. Cloud motions indicate the wind velocity at that level. Infrared pictures yield information on the surface temperature and on the temperature of the tops of clouds. Instruments that measure the radiation emitted in different wavelengths (radiometers) yield information about the vertical temperature and moisture structure of the atmosphere. Sensors that measure microwave energy provide data on the humidity and liquid water content. Although these quantitative data are not yet quite as accurate as radiosonde data, improvements are being made, and satellites are now providing much of the upper air data that are used for analysis and prediction.

1.4   AIR POLLUTION

Many types of air pollution have a variety of adverse effects on animal and plant life. Sulfur dioxide irritates the eyes, nose, and throat, damages the lungs, and aggravates asthma and bronchitis. When dissolved in precipitation, it forms sulfuric acid and can produce pH values as low as 3.0, which is acidic enough to dissolve marble statues and some fabrics. Particulates such as dust restrict visibility, dirty surfaces, and reduce the amount of sunlight reaching the ground. Oxidants, including nitrogen dioxide and ozone, irritate the eyes, aggravate lung diseases, and perhaps accelerate the aging process, as well as cause disintegration of many materials. Carbon monoxide in weak concentrations impairs the oxygen-transport function in the body and increases the general mortality rates; in large concentrations it can be fatal. Lead from automobile exhausts accumulates in the body, adversely affecting the brain, kidneys, and nervous system.

Air pollutants may be classified according to whether they are inert or reactive. Inert pollutants are those that react very slowly or not at all with other chemicals or gases in the atmosphere. Inert pollutants are often considered as passive substances because they are transported by the wind and diffused by turbulent eddies without changing their composition. Reactive pollutants, on the other hand, undergo chemical transformation as they encounter other substances or are altered by solar radiation. Pollutants may also be classified as gases or **aerosols**, which are solid or liquid particles, dispersed in the atmosphere. Major gaseous pollutants include ozone, sulfur dioxide, carbon monoxide, and oxides of nitrogen. Aerosols include smoke particles (carbon), bacteria, pollen, salt particles, and many types of dust. When averaged over the entire Earth, natural processes account for approximately 90 percent of the total particulate matter suspended in the air.

Concentrations of aerosols are usually expressed as the mass of the pollutant per unit volume of air, e.g., in units of micrograms per cubic meter. The concentration of gases may also be given in terms of mass per unit volume; however, gas concentrations are frequently given in terms of volume concentrations, i.e., the number of pollutant molecules per million molecules of air or parts per million (ppm). The relation between mass and volume concentrations at typical sea-level temperatures and pressures is

$$1 \, \mu g/m^3 = \left(\frac{24.5}{M} \right) ppm \tag{1.6}$$

where M is the molecular weight of the gas.

Aerosols occur over a large range of sizes (see Figure 1.3). Their formation occurs in two principal ways, *disintegration* of material and *agglomeration* of molecules. The formation of fine dust particles by the action of wind blowing over dry soil and the production of salt crystals by the breaking of bubbles on the sea surface and subsequent evaporation of the water are examples of the disintegration process. An example of aerosol formation by agglomeration is the production of water drops by condensation. Another example is the formation of soot by carbon molecules that combine after escaping combustion when oil or coal is partially burned.

The chemical composition of aerosols varies widely depending upon nearby sources. Common constituents, with typical concentrations of 10^{-6} grams per cubic meter, include sulfate (SO_4), nitrate (NO_3), chloride (CI), and ammonium (NH_4) ions. Sulfate aerosols with diameters of about 1 μm are the principal ingredients of large-scale pollution episodes which can cover several states and reduce visibilities in a murky haze to several kilometers or less. Sulfate ions are also ingredients of sulfurous smog (the type of smog associated with the infamous "London fogs" which plagued Great Britain from the thirteenth century until the early 1950s, when controls on the burning of sulfurous coal were imposed). On the other hand, photochemical smogs, produced when sunlight causes chemical reactions between oxides of nitrogen and organic compounds, contain large concentrations of nitrate ions.

Aerosols are removed by settling to the surface (*sedimentation*), *adherence* to solid objects after impact, and by wet processes of *rainout* and *washout*. Rainout refers to the incorporation of particles into cloud drops or ice crystals when the particles serve as cloud nuclei (see Section 2.1). Washout is the removal of particles by precipitation falling through polluted air.

Sulfur compounds are among the most prevalent and harmful pollutants. Sulfur dioxide (SO_2) was a major component of the London smog episode of December 1952, which caused an estimated 4000 deaths. Fuels such as oil or coal which contain sulfur produce sulfur dioxide (SO_2) when burned. Copper, zinc, or lead ores often contain sulfur, which is separated from the metal by roasting. Sulfur dioxide may be dissolved in water and oxidized, forming sulfuric acid (H_2SO_4), which is the primary cause of **acid rain**. Because SO_2 can exist for several days, it can be carried by the winds hundreds of kilometers from the source, producing regional air-pollution episodes.

Although its occurrence was recognized at least 250 years ago, acid rain is one of the most discussed and controversial air-pollution problems. The probable adverse effects of acid rain, as well as the deposition of dry acidic materials on the surface, include fish kills in acidified lakes (particularly in the northeastern United States and southeastern Canada), interference with vegetative and soil processes, corrosion of materials (such as limestone statues and buildings), and contamination of drinking water (by leaching of toxic metals by acidic water).

Figure 1.11 shows the average pH of rainfall over the United States over a four-year period. The **pH** is a measure of the acidity or alkalinity of a substance. A pH of 7.0 represents a neutral substance, such as distilled water. Values lower than 7.0 represent increasing acidity; values higher than 7.0 indicate alkalinity. A substance with a pH of 5.0 is ten times more acid than one with a pH of 6.0. For reference, vinegar has a pH of about 3.2, lemon juice 2.2, and battery acid 1.0. Because of the presence of carbon dioxide in the air, even unpolluted rain is slightly acidic, with a pH of 5.6. When the pH of water in lakes or streams decreases below about 5.0, most fish species die.

One of the controversial aspects of the acid rain problem is that the pollutants that form acidic precipitation (sulfur dioxide and oxides of nitrogen) are carried hundreds and perhaps thousands of kilometers downwind from their source, affecting remote areas that have few local sources of these materials. Figure 1.11 indicates that the most acidic precipitation in the United States occurs in the northeastern quarter of the country, which is downwind from the heavily polluting industrial plants in the Ohio River Valley. As a result, there have been many public calls from the affected areas to implement expensive controls of emissions at their source.

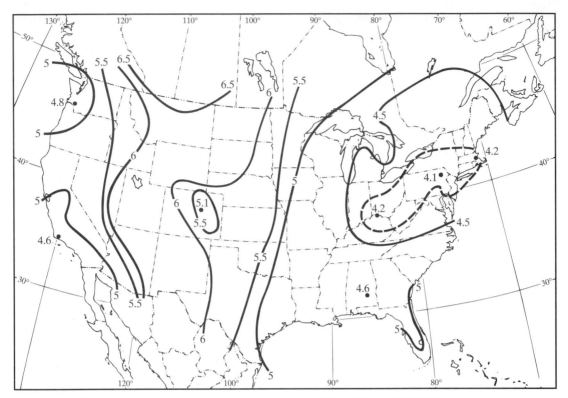

Figure 1.11   Average pH of precipitation over the United States, 1976-1979.

Oxides of nitrogen, consisting of nitric oxide (NO) and nitrogen dioxide (NO_2), are important ingredients in the formation of photochemical smog (the type found in Los Angeles). When heated by combustion engines such as the automobile, atmospheric nitrogen (N_2) is oxidized to NO, which may then be oxidized into the brown gas NO_2. When exposed to the short wavelengths of sunlight, the NO_2 is broken up into NO and free oxygen O. This atomic oxygen then combines with molecular oxygen O_2 to form ozone O_3, an irritating gas that causes rapid disintegration of materials such as rubber or plastic. Many additional reactions can occur between the ozone, oxides of nitrogen, and reactive hydrocarbons to produce the complicated mixture of photochemical smog.

Effect of Meteorology on Air Pollution

The two most important aspects of meteorology that affect air-pollutant concentrations are **wind speed** and **stability**. For a single source of a give emission rate, the volume of air affected by the source increases linearly with the wind speed; hence, other factors being equal, the concentration from a single source varies inversely as the wind speed. The stability of the atmosphere also plays an important role by governing the intensity of turbulence, which mixes air pollutants vertically and hence dilutes the polluted mixture. Under *unstable* conditions, with warm air near the ground and colder air aloft (as usually occurs during sunny days), vertical mixing occurs readily, with dirty air carried aloft and cleaner air mixed downward. Under *stable* conditions, when the temperature decreases slowly with height or even increases with height (an *inversion*), vertical mixing is inhibited and pollutants tend to remain at a constant elevation in the atmosphere.

Under normal conditions, a layer of unstable air occurs next to the ground with stable air aloft. The depth of this unstable layer in contact with the ground is called the **mixing depth** (Figure 1.12) because vertical mixing occurs easily in this layer. Typical depths of the mixed layer vary from about 600 meters over Los Angeles to a kilometer or two over cities in the Midwest or East. Pollutant concentrations tend to be vertically constant in the mixed layer. Therefore, the greatest pollution concentrations occur with shallow mixed layers or when an inversion occurs at the surface and vertical mixing is weak everywhere.

The effects of stability and wind speed form the basis for a simple prediction model of air pollutant concentration downwind of a single source:

$$X = \frac{CQ}{U} \tag{1.7}$$

where X is the concentration in micrograms per cubic meter, Q is the source rate in micrograms per square meter per second, U is the wind speed, and C is a dimensionless number that depends on the stability. Values of C are determined experimentally and range from 600 for stable to 50 for unstable conditions.

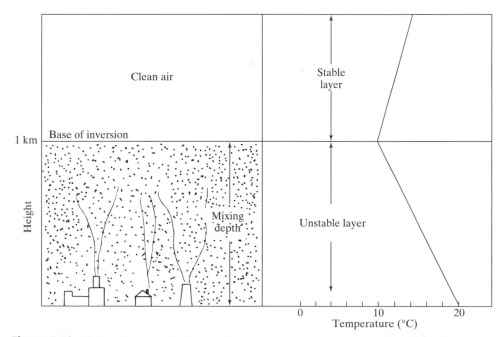

Figure 1.12 Pollutants mix vertically throughout mixed layer because the air is unstable. Above mixed layer, air is stable and vertical mixing is suppressed.

KEY TERMS

absolute humidity	homosphere	relative humidity
acid rain	inversion	saturation vapor pressure
aerosols	ionosphere	scales of atmospheric motions
carbon dioxide	latent heat	stability
density	mesopause	stratopause
dew-point temperature	mesosphere	stratosphere
diurnal	mixing depth	terminal velocity
equation of state	nuclei	thermosphere
exosphere	ozone	tropopause
extratropical	pH	troposphere
greenhouse gases	pressure	water vapor
heterosphere	radiosonde	wet-bulb temperature

PROBLEMS

1. If the average number of molecules per cubic centimeters is 25×10^{18} and the average density is 1.2×10^{-3} g/cm^3 at sea level, what is the average mass of an air molecule?

2. Modern jet aircraft often fly at an elevation of 10 km, at which level the pressure is about 300 mb. If the temperature is –40°C at this level, what is the density? (Express the answer as a fraction of the average surface density given above.)

3. Typical stratus clouds occur at an elevation of 1 km over the ground and have droplets of diameter 10^{-3} cm. How long would it take such a drop to reach the ground if there were no vertical air motions?

4. The average sea-level pressure is 1013 mb. What is this average expressed in inches of mercury?

5. If the global average sea-level pressure is 1013 mb and the average temperature is 10°C (283°K), what is the average density at sea level?

6. If you were to construct a scale model of the Earth and its atmosphere, starting with a 1-meter-diameter globe, how far from the surface would the following extend?
   a. Mt. Everest
   b. the level at which 99 percent of the atmosphere is found
   c. the level of the tropopause

7. Compute the height of a water barometer at a place where the atmosphere's pressure is 850 millibars, if the temperature of the barometer is 10°C.

8 Make a list of at least a dozen ways in which the atmosphere—its constituents and its motions—affect people and their activities.

9. As far as "weather" is concerned, which of the atmosphere's gases are most important? What role(s) does each play in the weather-making process?

10. The average atmospheric density at sea level is about 1.2×10^{-3} g/cm^3, and the average pressure is 1013 millibars. If the density were constant in the vertical, what would be the

depth of the atmosphere (at which point would $p = 0$)? What would be the depth of liquid water having the same pressure (1013 millibars) at the bottom?

11. If there is a fog in the morning and a temperature of 0°C, what will be the relative humidity in the midafternoon, when the temperature is 10°C, assuming that the actual vapor pressure remains unchanged?

12. In one of the worst pollution episodes in the United States (October 26-31, 1948), nearly half the population of Donora, Pa., became ill and approximately 20 people died from the high concentrations of pollutants. Particulate concentrations reached 4 mg/m^3 while SO_2 concentration exceeded 0.5 ppm. What was the SO_2 mass concentration in micrograms per cubic meter?

13. A moderate SO_2 emission rate is 1 g/km^2/s. Estimate the concentration of SO_2 in micrograms per cubic meter for the following conditions and compare the values with the peak SO_2 concentrations in the Donora, Pa. episode:
    a. unstable, wind speed = 5 m/s.
    b. stable, wind speed = 1 m/s.

14. A candy bar is labeled as containing 100 "calories." If you ate this candy bar, how many grams of water would have to be evaporated from your skin to cause your body to burn the 100 "calories" in the candy bar?

2
Clouds and Precipitation

2.1 CLOUDS

In the discussion on humidity, it was implied that as soon as the concentration of vapor begins to exceed the saturation value, which is dependent only on temperature, the excess vapor becomes liquid or solid. It is not quite that simple. There are surface intermolecular binding forces at the boundary of a liquid or solid that restrain the energetic molecules from escaping. (These binding forces produce what we commonly refer to as the "surface tension" of a liquid—the force that resists rupture of the surface and which we feel when we dive into a pool of water.) Whether or not a molecule will escape this surface restraining force depends on the molecule's speed when it reaches the surface. Although the *average* speed of molecules moving in a random fashion depends on the temperature, *individual* molecular speeds vary over a broad range of values. Some molecules may have sufficient momentum in the direction normal to the surface to escape; the higher the water temperature, the greater the percentage of molecules that will have the critical speed.

Even after some molecules have escaped from the bulk water, they continue to move about in all directions and over a range of speeds; inevitably, some may strike the surface, penetrate it, and be recaptured. In a given time interval, when move molecules are escaping than are being captured, we say there is **evaporation**; when the converse is true, there is **condensation**; and when the escape and capture rates are equal, there is an **equilibrium** state. The molecules of water vapor above the surface exert a pressure that depends on their number per unit volume and their temperature. Pressure exerted in the equilibrium state, with as many molecules condensing on the water surface as are evaporating from it, is called the **saturation vapor pressure**. The term **transpiration** is commonly used to describe evaporation of moisture from leaves and from other plant surfaces.

The equilibrium at saturation between evaporating and condensing molecules is altered either if the water surface is curved (because of surface tension effects) or if the water has impurities dissolved in it. Therefore, each water droplet has an equilibrium relative humidity slightly different from 100 percent. At environmental humidities greater than this equilibrium value the drop will grow; at humidities less than this value the drop will evaporate.

For condensation or **sublimation** to occur there must be a suitable surface available. **Dew** or **frost** forms easily on grass, soil, windows, etc., whenever the air temperature reaches the dew point or the frost point. But in the free atmosphere there are no such extensive surfaces.

In pure air (entirely gaseous, with no particles, water drops, or ice crystals), condensation or sublimation is extremely difficult to achieve because of the small probability of enough water-vapor molecules sticking together to form a embryonic drop or ice crystal. Fortunately, in the natural atmosphere there are numerous particles much larger than individual molecules that provide surfaces of **condensation nuclei** to which water molecules can adhere. It is around these that water droplets and ice crystals grow. (Cloud drops and raindrops are never "pure," despite the myth that they are, although the proportion of "foreign" particles to water is usually very small.) Certain types of particles, such as salts injected into the atmosphere from the sea, attract water molecules to their surfaces and are said to be **hygroscopic** nuclei, the effect of hygroscopic nuclei on the equilibrium relative humidity is called the solute effect. On these nuclei, condensation may actually begin well before the air becomes saturated. However, salt nuclei represent only a small number of the total particles suspended in the air; there are a large number of other types of particles, such as combustion products, meteoritic dust, and soil, that also serve as nuclei. These small particles, most of which have diameters of less than 1 micrometer (a thousandth of a millimeter) and are found in quantities of less than 1 micrometer (a thousandth of a millimeter) and are found in quantities of 10,000 or more per cubic centimeter, are so small that they remain suspended in the air for days at a time. (See Figure 1.3)

While hygroscopic particles allow water vapor to condense at relative humidity below 100 percent, the curvature of the tiny spherical droplets produces an effect that requires higher relative humidities for condensation than would be required for a flat surface of water. Thus a drop of pure water with a radius of 0.1 μm requires a relative humidity of 101 percent for equilibrium; at a humidity of 100 percent this drop would evaporate.

At temperatures below freezing, when the saturation vapor pressure is approached, water-vapor molecules may be converted directly to ice crystals, a process called sublimation. As with condensation, sublimation requires special tiny particles. The particles, called **sublimation nuclei**, are very rare in the atmosphere; hence nearly all clouds consist at first entirely of water drops and are produced by condensation rather than sublimation. Liquid water drops at temperatures below freezing are said to be supercooled.

Pure liquid water will not usually freeze without special nuclei called **freezing** nuclei. Certain materials are more effective as freezing nuclei at warmer temperatures than others. In fairly large volumes of water, the likelihood of having at least a few effective nuclei is quite high and freezing normally occurs at a temperature very close to 0°. But in very small droplets, it is possible to achieve a temperature as low as –40°C before freezing takes place. Indeed, in the atmosphere it is common for liquid water drops to exist in clouds at temperatures as low as –20°C. Large drops freeze at a higher temperature than small drops.

Fog and clouds are composed of liquid water and/or ice particles suspended in the air. There can be as many at 500–600 particles in each cubic centimeter, although normally there are fewer than half this number. However, the particles in nonprecipitating clouds are normally quite small (averaging about 0.01 millimeter in radius and rarely exceeding 0.1 millimeter), and so they fall to Earth very slowly. In calm air, a droplet having a radius of 0.01 cm falls at the rate of less than 50 cm/s (1 mi/h). This maximum fall velocity, known as the terminal velocity (Figure 1.3), is imposed by air resistance. Most clouds are formed when air is rising; so in practice, even in extremely weak upward air currents, drops of this size can be suspended in the atmosphere for many hours. In fact, clouds begin to precipitate only when some of the drops within them reach sufficient size to fall through the air with an appreciable velocity.

Cloud Types

Aside from observation of the internal makeup of clouds, the outward appearance of clouds is of significance to the meteorologist in interpreting the physical processes in the atmosphere, and it is often a harbinger of the weather to come. The basic clouds are identified on the basis of their form and the approximate height where they normally occur. The names of the basic clouds are composed of the following roots; *cirrus* (feathery or fibrous); *stratus* (stratified or in layers); *cumulus* (heaped up); *alto* (middle); and *nimbus* (rain). The ten basic clouds are grouped in four categories:

1. High (base over 7 km, or 23,000 ft, generally composed entirely of ice crystals): **cirrus** (Ci), **cirrostratus** (Cs), **cirrocumulus** (Cc)
2. Middle (2–7 km, 6500–23,000 ft): **altocumulus** (Ac), **altostratus** (As)
3. Low (below 2 km, or 6500 ft): **stratus** (St); **stratocumulus** (Sc), **nimbostratus** (Ns)
4. Clouds of vertical development (base usually below 2 km, or 6500 ft, but top can extend to great heights): **cumulus** (Cu), **cumulonimbus** (Cb)

Examples of these clouds appear in Figure 2.1. Some common adjectives applied to the basic names to further describe particular clouds are:

1. *Uncinus:* hook-shaped—applied to cirrus; often shaped like a comma
2. *Castellanus:* turreted—applied most often to cirrocumulus and altocumulus
3. *Lenticularis:* lens-shaped—applied mostly to cirrostratus, altocumulus, and stratocumulus; occurs where air currents are undulating sharply in the vertical, as sometimes occurs on the lee sides of mountains
4. *Fractus:* broken—applied only to stratus and cumulus

5. *Humilis:* lowly—poorly developed in the vertical; applied to cumulus

6. *Congestus:* crowded together in heaps, like a cauliflower—applied to cumulus

7. *Mamma (mammatus):* hanging protuberances, like udders, on the undersurface of a cloud

As will be shown in the next section, almost all clouds result from the rapid cooling of air where it ascends. In straitform clouds, the motion in the vertical is generally small (less than 50 centimeters or 1 mi/h) while in cumuliform clouds, the upward and downward velocities are much stronger (up to 30 m/s—70 mi/h—or more).

Fog is merely a cloud in which the observer is immersed. It is thus a suspension of small water droplets. When the fog is formed of tiny ice crystals, it is known as *ice fog*. Haze often precedes and follows fog. **Haze** is formed of very small droplets of wet hygroscopic particles. The air does not feel "wet" as it does in fog.

Smog originally meant a natural fog contaminated with pollution, i.e., a mixture of smoke and fog. It has recently come to mean any general urban air-pollution mixture in which visibility is significantly reduced, whether or not natural fog is actually present. Therefore, there are a variety of smogs, some produced through photochemical reactions between nitrogen oxides and hydrocarbons (both emitted primarily by automobile exhausts) and some produced by sulfurous fumes emitted by combustion of fuel oils and coal of high sulfur content. Because the chemical constituents of smogs are often hygroscopic, tiny water drops can exist at relative humidities as low as 80 percent. Actually, these haze droplets are often drops of sulfuric acid, with pH values as low as 3.0.

Figure 2.1a Condensation trail spreading horizontally (top), fibrous circus (center), patch of cirrostratus (left), and cumulus clouds (top of ridges, right). (Source: NCAR photograph.)

Figure 2.1b   Altocumulus with waves. (Source: Photograph by Richard Anthes.)

Figure 2.1c   Stratocululus layer with virga (precipitation falling from cloud but not reaching ground). (Source: NCAR photograph.)

Figure 2.1d   Stratocululus with cirrostratus clouds above. (Source: NCAR photograph.)

Formation of Different Cloud Types

Because of the abundance of cloud nuclei in the atmosphere, clouds will be formed whenever the actual vapor pressure exceeds the saturation value. Theoretically, this could occur by either increasing the actual vapor pressure or decreasing the saturation

Figure 2.1e   Stratus with base below the ridge top. (Source: Photograph by Richard Anthes.)

Figure 2.1f Marine fog layer from above. Cirrocumulus patches aloft. (Source: Photograph by Richard Anthes.)

Figure 2.1g Towering cumulus. (Source: NCAR photograph.)

Figure 2.1h   Cumulonimbus. (Source: Photograph by Robert Stone.)

vapor pressure. Since it is impossible to increase the actual vapor pressure past the saturation value by evaporation, the only practical alternative for forming clouds is to decrease the saturation value. Such a decrease is accomplished when the air temperature

Figure 2.1i   Mamma in lower troposphere. (Source: Photograph by Dennis Thomson.)

Figure 2.1j Orographic wave clouds composed mainly of ice crystals. (Source: Photograph by Henry Van de Boogaard.)

Figure 2.1k Kelvin-Helmholtz clouds. (Source: NCAR photograph.)

is lowered, because the saturation vapor pressure decreases rapidly with decreasing temperature (see Figure 1.9). The required drop in temperature can occur through mixing with colder air or by cooling of the air parcel.

The formation of clouds by mixing is illustrated in Figure 2.2, which shows a graph of the saturation mixing ratio over water at a pressure of 1000 mb. The **mixing ratio** is defined as the mass of water vapor per mass of dry air; the saturation mixing ratio is the value at saturation. Suppose there are two parcels of air at different temperature and mixing ratios, as indicated by points A and B in Figure 2.2. Because the mixing ratio of both parcels is less than the saturation value, clouds will not exist in either parcel. Consider what happens if the two parcels are mixed, however. For parcels of equal mass, the temperature and mixing ratio of the mixture will be the average of the original values in each parcel. This average of 0°C and 4.6 g/kg is shown by point D in Figure 2.2. Note that because the saturation mixing ratio curve is concave, the mixing ratio of the mixture exceeds the saturation mixing ratio at the temperature of the mixture. Thus a mixing of two unsaturated parcels of air has produced a parcel that is supersaturated, and one in which cloud formation would be very likely. An example of cloud formation by mixing is sea smoke (or steam fog), which is formed when cold air flows over warm water and mixes with the warm, moist air in contact with the water. Another example is the condensation trail (**contrail**), produced when hot, moist exhaust from jet engines mixes with cold environmental air. But perhaps the most familiar example of a cloud produced by mixing is the cloud formed when moist breath is exhaled on a cold day.

The cooling of extensive layers of air to the dewpoint is the most common mechanism for producing clouds. Cooling may occur by radiation, by conduction when air comes into contact with a colder surface, or through lifting.

Several mechanisms may cool the air near the Earth's surface sufficiently to form fog. **Radiation fog** is formed when radiative cooling of the Earth's surface and low-level air chill the air to below the dew point. Radiation fogs usually occur on calm clear nights when the relative humidity is high. Because the air cooled by radiation is dense, it flows downward into valleys where it forms pools of chilled air; thus radiation fog is most common in deep valleys and less common on the tops of hills and mountains.

Figure 2.2 Formation of a cloud by mixing parcels A and B.

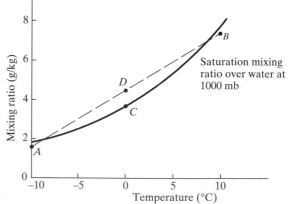

Advection fog is formed when warm, moist air is advected (moved) over a cooler surface. Advection fogs occur commonly on the west coast of the United States when moist Pacific air moves over the cold California current. They also occur frequently when warm, moist air is carried over snow or ice cover.

Upslope fogs occur when humid air is lifted over a hill, mountain or gradually sloping plain (such as occurs with an east wind blowing over the sloping Great Plains of the United States).

Because the mechanisms that produce fog are relatively slow processes and affect only a thin layer of air near the surface, fogs and stratus clouds usually do not produce precipitation. However some produce a light drizzle which can be important to some ecosystems where fog is a frequent occurrence (such as the west coast of the United States).

The mechanism that produces the most rapid rate of cooling and, therefore, almost all precipitation, is the cooling of rising air by expansion. As discussed in Section 4.4, dry air cools 9.8°C per km of rise. With condensation and the accompanying release of latent heat, the rate of cooling is less but still exceeds 6°C/km. Therefore, a parcel of air in a thunderstorm updraft can cool 40°C in 15 min as it is lifted 8 km or more. Even the large nonconvective cloud sheets associated with middle-latitude storm systems can ascend over 4 km in one day, producing cooling of 30 to 40°C during this time. Vast sheets of cirrostratus, altostratus, and nimbostratus are produced in the region around cyclones (low-pressure systems) in which the air is moving upward.

When the air is stable (see Section 4.5), small-scale vertical motions are suppressed and clouds take on a smooth, uniform layer appearance. When the atmosphere is less stable, small-scale saturated parcels of air may become buoyant. As many of these buoyant elements rise, a turreted appearance to the clouds evolves. Thus *castellanus* clouds indicate relatively unstable layers of air.

Waves of alternating upward and downward motion often exist in the atmosphere. If the air is close to saturation, the small amount of lifting can produce a limited region of condensation, and a wave cloud results (Figure 2.3). Lifting of air over mountains often produces a similar *lenticularis* cloud.

Under appropriate conditions of atmospheric stability and **wind shear** (variation of wind with height), waves in the atmosphere call **Kelvin-Helmholtz** waves may occur. These waves, which are very similar to waves in the ocean, may break as illustrated by the clouds in Figure 2.1 (k).

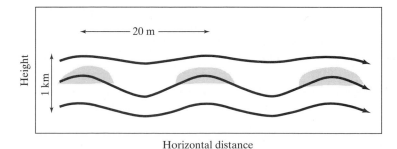

Figure 2.3 Wave clouds formed when air in the rising portion of the wave is cooled below the dewpoint, when air comes in contact with a colder surface, or through lifting.

Evaporation

Evaporation from a wet surface occurs whenever the water vapor pressure in the free air is less than the saturation pressure at the surface, i.e., the air is not saturated. The factors that affect the rate of evaporation from a surface are: (1) the temperature of the water surface (since high temperature implies a high saturation vapor pressure), (2) the actual and saturation vapor pressures in the free air near the surface, and (3) the effectiveness of diffusion of vapor away from the surface. As we saw in our discussion of mixing by the wind, the diffusion of water vapor is greatest when the wind is strong and gusty.

Wind has another effect that enhances the evaporation over large water bodies. Spray ejected into the air from wind-generated waves is composed of very small droplets (diameter less than 0.05 millimeter), which evaporate rapidly in the air. When the wind speed is greater than 50 km/h (35 mi/h), the rate of evaporation of spray may actually exceed the rate from the surface of the water body.

Direct measurement of the worldwide evaporation rate is not practical, and estimates are based on computations from those theoretical relationships given previously. Evaporation pans (shallow circular pans filled with water, with a scale to indicate the changes in depth) are often used to "measure" the evaporation from dams and reservoirs and over cultivated fields. However, the evaporative loss from such pans is almost invariably higher than the true evaporation because of differences between pan and natural surface properties and the difference in "fetch" (i.e., the distance travelled by the air in traversing the surface).

Most of the world's evaporation occurs over the oceans, and it is especially high throughout the year over the tropical latitudes between 5 degrees and 30 degrees (4–7) millimeters per day) and over the western parts of oceans in middle latitudes during the cooler half of the year (4–9 millimeters per day), when dry cool air from the continents sweeps over such warm ocean currents as the Kuroshio and the Gulf Stream.

2.2 PRECIPITATION

Raindrops are usually between one and several millimeters in diameter. Since the average drop diameter in a nonprecipitating cloud is about 0.02 millimeter, this means that many cloud drops have increased their volume by a factor of 1,000,000 by the time they fall out of a cloud. Such growth cannot be explained merely by further condensation of water vapor on existing water particles because the process becomes very slow for larger drops. Small particles must therefore unite to form large particles. How this is accomplished is still not completely settled, but the most probably mechanisms are the following:

Collision and Coalescence of Particles. The drops formed in a cloud are not all of the same size. Due to differences in the rate at which condensation proceeds in different parts of a cloud—sometimes separated by very small distances—the largest drops in the cloud may have diameters several times greater than the smallest. As the air swirls about, the larger drops, because of their greater mass, have more inertia than the smaller ones. These large drops tend, therefore, not to follow exactly the same path as the small ones, and as a result, collide and often coalesce (combine) with the small ones. Repeated collision and coalescence by a drop may cause it to grow so large that it splinters into several drops, which in turn, grow by collision and coalescence, thus producing a "chain reaction" of raindrop growth.

Growth of Ice Crystals.   Frequently, especially in the middle latitudes, the upper-most layers of clouds will be composed of ice crystals, while the lower layers contain **supercooled** (temperature below 0°C) liquid drops. Through stirring within the cloud, or because of different fall velocities of the particles, ice crystals and supercooled drops become mixed. At the same temperature, the saturation vapor pressure over a liquid surface is greater than that over an ice surface (see insert of Figure 1.9), and the ice crystals will therefore grow at the expense of the water drops. This growth mechanism, which is called the **Bergeron process** after the Swedish meterologist who first suggested it, is believed to be quite important in the *initiation* of precipitation, although further growth most likely involves the collision-coalescence of particles described in the previous paragraph.

The importance of coexistence of ice crystals and supercooled water drops in the initiation of precipitation forms the basis for many modern cloud-seeding experiments. In a cloud that contains few or no ice crystals, dry ice introduced into the cloud may cool enough drops to their freezing point to produce crystals for later growth by the Bergeron process. Supercooled droplets can also be induced to freeze through injection of certain types of materials, such as silver iodide, apparently because the crystal structure of these materials is very similar to that of ice (Figure 2.4).

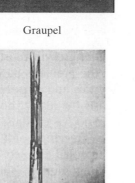

Graupel

Dendretic crystal

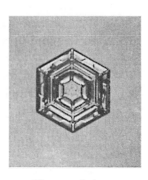

Hexagonal plate

Needle

Sector-like crystal

Sheath-like crystal

Figure 2.4   Some forms of snow crystals. (Source: C. Magono, Hokkaido University, Japan.)

Precipitation Types

The only difference between *drizzle* and *rain* is the size of the water droplets. The diameter of the former is generally less than 0.5 millimeter. The principal frozen forms of precipitation are

1. **freezing rain** or **drizzle**: Rain or drizzle that freezes on impact with the ground or objects
2. **sleet** (ice pellets): Small ice particles or pellets that originated as rain but froze as they traversed a cold layer near the ground
3. **snow:** white or translucent ice crystals, chiefly in complex branched hexagonal form and often agglomerated into snowflakes (Figures 2.4 and 2.5).

Figure 2.5   Denver, Colorado blizzard of 1982. (Source: NCAR photograph.)

4.   **hail:** small balls or chunks of ice with a diameter of 5–50 millimeters (0.2–2 inches) or more that fall from cumulonimbus clouds. These destructive stones are formed by the successive accretion of water drops around a small kernel of ice falling through a thick cloud; as each drop is frozen onto the nucleus, it may form a new shell, so that many hailstones acquire an onionlike cross section (Figure 2.6).

Measurement of Precipitation

For practical purposes of water supply, meterologists are concerned with measuring the amount of water reaching the Earth's surface. This is done by sampling the depth of water that would cover the surface if the water did not run off or filter into the soil. A rain gauge is merely a collection pail with a ruler to measure the depth of water. The depth is usually measured in increments of a hundredth of an inch or a millimeter. Solid forms of precipitation are melted and the equivalent liquid depth recorded. In the case of snow, which may remain on the ground for a long period of time and thus serve as a natural water reservoir, the snow depth is of interest. Normally, the ratio of snow depth to liquid equivalent is about 10 to 1, although sometimes it is as much as 30 to 1.

Precipitation amounts vary considerably from place to place, even during a single storm, so that the problem of adequate sampling over an area is a serious one. In most places in the world, not more than one 20-cm diameter rain gauge is installed in every 200 square km. In this case, the fraction of the 200 square km area actually sampled by the rain gauge is given by the ratio of the area of the rain gauge (314 cm^2) to the area of the region $(200 \times 10^5)cm^2$ or approximately $1:10^{12}$. This is somewhat like taking a single hair from one Californian's head to characterize the hair of everyone in the state. In mountainous areas especially, amounts may vary by a factor of two or three in a distance of less than 20 km. (Along the northeast slopes of Hawaii, the annual rainfall varies from 40 cm to over 750 cm in a distance of about 25 km.) Interpretation of measured amounts must be done with considerable care. Even at the

a.

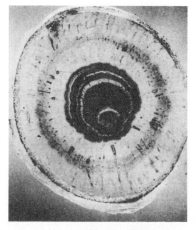

b.

Figure 2.6   **a.** Hailstone which fell over Coffeyville, Kansas, September 3, 1970. (Source: NCAR photograph.) **b.** Cross section of a hailstone. (Source: Roland List.)

same point, annual amounts vary greatly, especially in semiarid climates; the year-to-year amounts can easily fluctuate by 50 percent or more of the long-term average annual precipitation. For this reason, claims by rainmakers that they have increased the rainfall by some precise figure, like 10.4 percent, should be viewed with a great deal of skepticism.

The Hydrologic Cycle

Humans have always been highly dependent on water supply. Ancient civilizations have withered or prospered as precipitation patterns have shifted over the centuries. Today, the increasing growth and concentration of population and of industry are increasing the demands on an essentially fixed world water supply, and we have become more concerned than ever with the need to carefully budget our water resources.

One of the hydrologists's tasks is to study the distribution of the world's water supply. As shown in Table 2.1, the Earth's total water supply is about 1.4×10^9 km^3 and is contained mostly in the oceans (96.5 percent). Glaciers and permanent snow cover contain about 1.7 percent and ground water contain another 1.7 percent. Most of the global water supply is saline; the percentage of the total amount of water that is fresh is a mere 2.5 percent. The atmosphere, although containing only about 0.001 percent of the total global water on the average, provides the vital "hydrologic link" between oceans and land.

TABLE 2.1   Water reserves on the Earth (from Gleick, 1993)

	Volume (10^3 km^3)	Percentage of global reserves	
		Of total water	Of fresh water
World ocean	1,338,000	96.5	—
Ground water	23,400	1.7	—
Fresh water	10,530	0.76	30.1
Soil moisture	16.5	0.001	0.05
Glaciers and permanent snow cover	24,064	1.74	68.7
Antarctic	21,600	1.56	61.7
Greenland	2,340	0.17	6.68
Arctic islands	83.5	0.006	0.24
Mountainous regions	40.6	0.003	0.12
Ground ice/permafrost	300	0.022	0.86
Water reserves in lakes	176.4	0.013	—
Fresh	91	0.007	0.26
Saline	85.4	0.006	—
Swamp water	11.47	0.0008	0.03
River flows	2.12	0.0002	0.006
Biological water	1.12	0.0001	0.003
Atmospheric water	12.9	0.001	0.04
Total water reserves	1,385,984	100	—
Total fresh water reserves	35,029	2.53	100

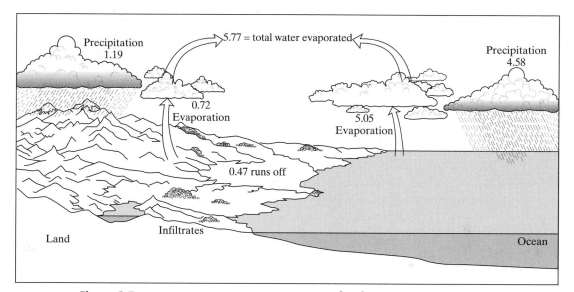

Figure 2.7 Earth's annual water balance, units are 10^5 km^3 of water per year. Huge quantities of water are cycled through the atmosphere each year. Notice that, on the continents, precipitation exceeds evaporation and that the reverse is true for the oceans. Because the level of the world's ocean is not dropping, runoff from the continents must balance the deficit of the oceans.

The annual global hydrologic cycle is illustrated in Figure 2.7. For this discussion, consider a unit of water volume equal to 10^5 cubic kilometers (km^3). Each year, on the average, 1.19 units of water falls over land areas as precipitation. Of this total, 0.72 units is evaporated, leaving 0.47 units to run off into the oceans. Over the oceans, the precipitation is 4.58 units, which is exceeded by an evaporation rate of 5.05 units. The difference between the evaporation and precipitation over the oceans is made up by the runoff from the land.

Even if all of the water vapor carried in the atmosphere at any moment could be precipitated, an average depth of only about 2.5 centimeters of liquid water would result. But new water is constantly being added through evaporation [at the rate of 5.8×10^5 km^3 per year, more than 85 percent coming from the seas], as well as being extracted through precipitation. Within the moving currents of air, the amount of water passing over an area during a period of time can be considerable. For example, even over arid Arizona, the atmosphere carries as much water in a single week in July as flows in the Colorado River during an entire year. Actually, even during heavy precipitation, only a small percentage of the total water in a vertical air column is precipitated; the intense inflow of air into a storm (as in a hurricane) more than compensates for the low rate of moisture extraction. On the average, only about one-tenth of the total water vapor that passes over the U.S. actually precipitates.

Although the worldwide precipitation rate of 113 cm/yr must equal the global evaporation rate, precipitation and evaporation are far from uniform throughout the world. Of the total water supply from precipitation, 79 percent falls over the oceans. The average annual amounts falling on continents vary enormously—from near zero centimeters to more than 700 centimeters (276 inches), with some points receiving as much as 1100 centimeters. In the conterminous United States, the average annual rainfall is about 75 centimeters (30 inches), but ranges from as little as 5 centimeters (2 inches) in the deserts of the Southwest to almost 380 centimeters (150 inches) in the Cascade Mountains of Washington and Oregon.

Of the precipitation that falls over continents, about 60 percent is returned to the atmosphere through evaporation from soil, lakes, rivers, animals, and vegetation; the rest runs off into rivers (or infiltrates into the soil), eventually emptying into the oceans. The average evaporation rate from oceans is almost double that over land (485 mm/yr), and since the area of the oceans is almost triple that of land, the oceans account for more than 85 percent of the worldwide evaporation.

Hydrology deals with all factors that affect water supply over land—precipitation, storage in soil and underground reservoirs, runoff, evaporation, and the return of moisture to the atmosphere and oceans. In the United States, a network of more than 13,000 precipitation gauges measures rain and snow; more than 300 evaporation pans are used to estimate evaporation. Storage of water in snowfields is obtained by measurements of snow depth. The flow rate and depth of rivers are also measured. From such information, the hydrologist attempts to anticipate destructive floods, drought, and future water supply, and to design reservoirs, sewer systems, bridges, etc.

Because snowfall builds up over the winter season and is released relatively rapidly when spring arrives, flooding may result. Also, hydrologists need to estimate the runoff available for agricultural purposes. Satellites have been used to estimate snowpack over large inaccessible regions, such as the mountainous areas of the western United States. Figure 2.8(a) shows NASA's Landsat satellite's view of the snow cover over the Sierra Nevadas in February of a normal year (1975). Figure 2.8(b) shows the same area during the drought year of 1976–1977.

Snowcover differences between a near normal runoff season (1975) and a drought year (1977) in the Sierra Nevada Mountains near Lake Tahoe, California as observed by LANDSAT.

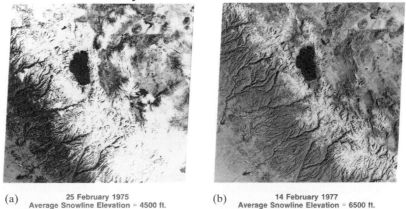

(a)   25 February 1975
      Average Snowline Elevation = 4500 ft.

(b)   14 February 1977
      Average Snowline Elevation = 6500 ft.

Figure 2.8   Landsat views of snow cover on the Sierra Nevada during (a) a normal runoff season (1975) and (b) a drought year (1977). (Source: NASA photograph.)

KEY TERMS

altocumulus

altostratus

Bergeron process

cirrocumulus

cirrostratus

cirrus

coalescence

collision

condensation

condensation nuclei

contrail

cumulonimbus

cumulus

dew

equilibrium

evaporation

fog

freezing rain

frost

hail

haze

hydrologic cycle

hygroscopic

nimbostratus

saturation vapor pressure

sleet

smog

stratocumulus

stratus

sublimation

sublimation nuclei

supercooled

transpiration

PROBLEMS

1. The evaporation-precipitation cycle behaves like an enormous desalinization plant. If we wanted to double the natural desalinization rate, how much energy would be required each year? How does this compare with the U.S. power production?

2. How "pure" is the typical raindrop; i.e., what is the ratio of solute mass (condensation nucleus) to water mass? (Assume that the typical nucleus has a diameter of 10^{-4} millimeters and a density of 2.5 times that of water and consult Figure 1.3 for average size of raindrop.)

3. The atmosphere contains about 2.5 centimeters of precipitable water and precipitates an average of 100 cm/yr. What must be the mean "residence time" of water in the atmosphere?

4. A typical cloud water content is 1 g/m^3. If a cloud with this amount of water were 3 km thick and all the cloud water were precipitated out, how deep would the precipitation be?

5. Explain why there is some truth to the proverb "It's too cold to snow."

3

The Atmosphere's Energy

3.1 ENERGY BUDGETS AND HEAT ENGINES

A convenient way to examine the workings of the atmosphere is through an energy budget. The law of the conservation of energy requires that we account for all of the energy received by the Earth, so by looking at all forms of energy and transformations we have a guide to atmospheric phenomena. This is similar to following in detail what happens to the fuel energy provided in an engine, thereby ending up with a fairly good picture of the operation of the engine.

Practically all (99.98 percent) of the energy that reaches the Earth comes from the *sun*. Intercepted first by the atmosphere, a small part is directly absorbed, particularly by certain gases such as ozone and water vapor. Some of the energy is reflected back to space by the atmosphere, its clouds, and the Earth's surface. Some of the sun's radiant energy is absorbed by the Earth's surface. Transfers of energy between the Earth's surface and the atmosphere occur in a variety of ways, such as radiation, conduction, evaporation, and convection. **Kinetic energy** (energy of air in motion or wind energy) results from differences in temperature within the atmosphere in much the same way that a heat engine converts differences in heat levels between the inside and outside of the expansion chamber to the motion of the piston. And, finally, turbulence and friction are constantly bleeding off some of the energy of motion, converting it to heat. The combination of these many processes produces the complex atmospheric phenomena that determine our weather.

Transfer of Heat Energy

Heat energy can be transmitted from one place to another by conduction, convection, and radiation.

Conduction is the process by which heat energy is transmitted through a substance by point-to-point contact of neighboring molecules, even though the molecules do not leave their mean positions. Solid substances, especially metals, are usually good conductors of heat, but fluids such as air and water are relatively poor conductors. Heat conduction in air is so slow that it is of little importance in transmitting heat within the air itself; however, it is significant in the exchange of heat between the Earth's surface and the air in contact with it.

Convection transmits heat by transporting groups of molecules from place to place within a substance. Thus, convection occurs in substances in which the molecules are free to move about; i.e., in fluids. The convective motions that carry heat from one point to another within a fluid arise because the warmer portions of the fluid are less dense than the surroundings and therefore rise, while the cooler portions are more dense and therefore sink. A circulation of fluid motion is thus established between the warm and cool regions. Much more will be said about convection in later chapters because it is an important heat transfer process in the atmosphere.

Radiation is the transfer of heat energy without the involvement of a physical substance in the transmission. Heat may therefore be transmitted through a vacuum, and if the radiation takes place through a completely transparent medium, the medium itself is not heated or otherwise affected. The transfer of the heat energy from the sun to the Earth is by means of the radiative process. Earth also loses its heat energy to outer space in the same way: by radiative transfer.

3.2 SOLAR ENERGY

Since the atmospheric "heat engine" is powered by solar energy, we start our discussion with the source of the "fuel," the sun. The sun is not an unusual star, either in brilliance or in size. A slowly rotating body of hot (several million degrees Celsius) very dense gas, with a diameter of about 1,400,000 kilometers (870,000 mi), it is surrounded by a very tenuous atmosphere that extends several solar diameters from the surface. It generates a tremendous amount of heat (about 5.6×10^{27} calories are radiated every minute, or 3.9×10^{23} kilowatts of power), but the Earth intercepts less than one part in two billion of this total. Measurements made on the Earth indicate that the rate at which energy impinges on a surface perpendicular to the sun's rays at the mean sun-Earth distance is about 2 cal/cm^2/min (1370 watt/m^2). This value is known as the solar constant, although no one is certain exactly how "constant" the output of the sun is. However, variations of the total energy output over the past decade are certainly within ± 0.1 percent. The intermittent outbursts of small particles and very short radiation, associated with disturbances on the sun, are largely absorbed in the outermost layers of the atmosphere.

According to Bethe's theory, the energy radiated from the sun is created through a complex thermonuclear reaction that converts protons (hydrogen nuclei) to alpha particles (helium nuclei). In the process, mass is converted to energy (in accordance with Einstein's familiar relationship, $E = mc^2$). The gravitational contraction of the enormous mass of the sun produces the temperature needed to make such a reaction possible. Although the sun is now converting mass to energy at the rate of about 4×10^6 tons per second, judging from the number of protons still available, the sun should continue shining for about 10^{11} years.

The radiant energy transmitted by the sun covers a broad range of the **electromagnetic spectrum**, from the very short gamma rays and X-rays to radio wavelengths (Figure 3.1). The rate at which energy is emitted from each square centimeter of surface as a function of wavelength is very much like that for an ideal or **black body** at 6000 K, shown in Figure 3.2. Eighty percent of the solar energy falls in the visible (0.38–0.72 μm*) and near infrared (0.72–1.5 μm) portions of the spectrum, with the peak energy occurring at a wavelength of approximately 0.5 μm, which is blue. In contrast, a body of 300 K (27°C, which is about a dozen degrees warmer than the mean temperature of the air near the Earth's surface) radiates energy at a rate of $1/160000$ that of a body at 6000 K, almost entirely in the far infrared, with a maximum emission near 10 μm wavelength.

The above properties of the radiation spectrum follow two important laws, the Stefan-Boltzmann law and Wien's law. The Stefan-Boltzmann law relates the total amount of energy (E) radiated by a blackbody (perfect radiator) to the fourth power of the absolute temperature

---

*Wavelengths are usually given in micrometers, abbreviated μm, or in angstrom units, abbreviated A. One micrometer is 10^{-4} cm, or 0.0001 cm, and 1 Å = 10^{-8} cm, or 0.00000001 cm.

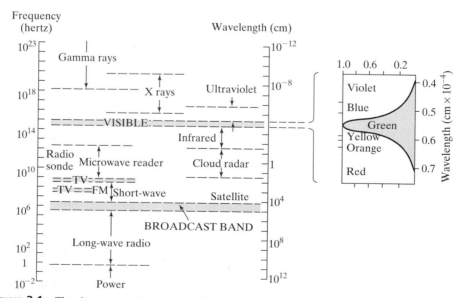

Figure 3.1 The electromagnetic spectrum. The curve at the right gives the relative sensitivity of the human eye to the electromagnetic radiation (maximum sensitivity at 0.555×10^{-4} cm wavelength).

Figure 3.2 Black-body emission of
(a) a hot body such as the Sun, and (b) a
cool body such as Earth. (In comparing
the curves, note that the vertical scale of
(a) is 100,000 times that of (b).)

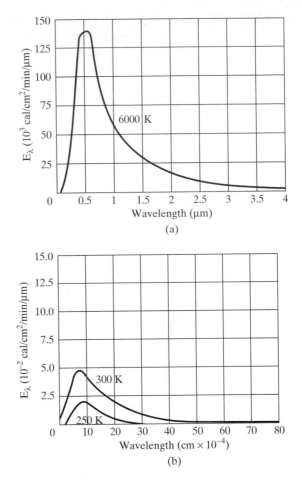

$$E = \sigma T^4 \tag{3.1}$$

where σ is the Stefan Boltzmann constant equal to 5.67×10^{-8} Wm^{-2}K^{-4}. The wavelength at which a body emits most intensely is inversely proportional to the temperature, and is given by Wien's law

$$\lambda_{max} = C/T \tag{3.2}$$

where C is 2898 μm K. Wien's law explains why the sun, with a temperature of 6000K
emits its maximum energy at a wavelength of about 0.5 μm while the Earth, with a
mean temperature of about 285 K emits its maximum energy at about 10 μm. Thus,
hot bodies not only radiate much more energy than cold bodies, but they do it at
shorter wavelengths. "Red hot" is not as hot as "blue hot."

Solar Energy in the Atmosphere

When the energy in the form of electromagnetic waves encounters the Earth's atmosphere, some of the rays pass undisturbed, some are absorbed by the atmosphere, and the rest are turned back. We shall examine the ways in which all three of these occur.

Absorption.   Oxygen, ozone, water vapor, carbon dioxide, and dust particles are the most significant **absorbers** of the **short-wave** radiation from the hot sun and the **long-wave** radiation of the cool Earth. The gases are *selective* absorbers, meaning that they absorb strongly in some wavelengths, weakly in others, and hardly at all in still others. The very short ultraviolet (less than 0.20 μm) radiation of the sun is absorbed as it encounters and splits molecular oxygen into two atoms in the upper levels of the atmosphere. Ozone, formed by the combination of O and O_2 effectively absorbs ultraviolet light of longer wavelengths—those between 0.22 and 0.29 μm. Figure 3.3 illustrates the relative effectiveness of oxygen and ozone as absorbers. Absorptivity is the fractional part of incident radiation that is absorbed. It can be seen that oxygen

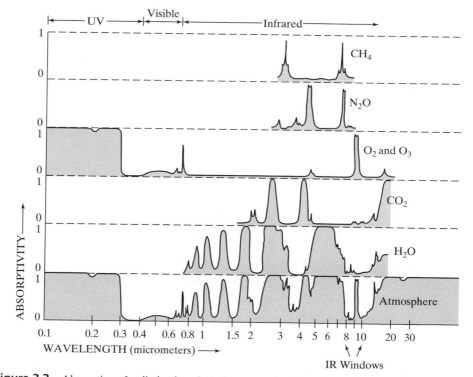

Figure 3.3   Absorption of radiation by selected gases as a function of wavelength. Absorptivity is the function of the radiation absorbed and ranges from 0 to 1 (0% to 100% absorption). Absorptivity is very low or near zero to atmospheric windows. Note the infrared windows near 8 and 10 micrometers (Source: R.G. Feagle and J. Businger, *An Introduction to Atmospheric Physics*, New York: Academic Press, 1980 and Morgan and Moran, 4th Ed.)

and ozone absorb almost 100 percent of all radiation at wavelengths less than 0.29 μm. For this reason, only a minute portion of the sun's ultraviolet radiation penetrates to the lower levels of the atmosphere. In the longer wavelengths neither of these gases absorbs very much energy, except for a narrow band (near 9.6 μm) in the infrared. About 2 percent of the sun's total radiation received on earth is depleted by ozone.

Water vapor is a significant absorber of radiation. Its complicated absorptivity characteristics are illuminated in Figure 3.3. Although not effective at wavelengths below 0.8 μm, where most of the *solar* radiation exists, it absorbs (strongly between 5 and 7 μm and moderately well beyond 15 μm) wavelengths at which the Earth and its atmosphere emit much of their energy (see Figure 3.2).

In summary, the clear atmosphere is essentially transparent between 0.3 and 0.8 μm, where most of the solar (short-wave) radiation occurs. But between 0.8 and 20 μm, where much of the terrestrial (long-wave) radiation is emitted, there are several bands of moderate absorptivity by water vapor, carbon dioxide, and other trace gases.

Scattering. The atmosphere is composed of many, many discrete particles—gas molecules, dust, water droplets, etc.—but the empty space between particles is actually greater than the volume occupied by the particles. Each particle acts as an obstacle in the path of radiant energy (e.g., light waves) traveling through the atmosphere, much as rocks in a lake impede the progress of ripples in the water. The wave fronts are deformed by these obstacles into a pattern that makes it appear that the rays emanate from the obstacles. Thus, radiant energy propagating in a single direction is dispersed in all directions as it encounters each particle in its path. This dispersion of the energy is called **scattering**.

The scattering effectiveness of a particle depends on its volume. For particles the size of gas molecules, the amount of scattering is much greater for the short wavelengths of light (blue) than for the long waves (red). It is for this reason that the white light of the sun and the moon becomes yellow or red on the horizon, while the sky, which is lit by scattered lights, is blue. Astronauts observe, as they ascend through the atmosphere, that the sky becomes darker, finally becoming black, as the density of the scattering particles decreases.

When the atmosphere contains many large dust particles or minute water droplets (haze), scattering is no longer very selective in terms of wavelength. The long waves are scattered almost as much as the short waves, and the resulting sky color becomes less bluish and more white or milky. In fact, the blueness of a cloudless sky is an indication of its "purity," i.e., how free it is of smoke, dust, and haze.

On the average, about 12 percent of the sun's radiation striking the Earth's atmosphere is scattered; half of the scattered radiation is lost to space.

Reflection. The radiant energy from the sun encounters still another obstacle before it reaches the surface of the Earth—clouds. Most clouds are very good reflectors but poor absorbers of radiant energy. The **reflection** by clouds depends primarily on their thickness, but also to some extent on the nature of the cloud particles (i.e., whether ice or water) and the size of these particles. The reflectivity of clouds varies from less than 25 percent to more than 80 percent, depending on the cloud thickness. Clouds absorb very little of the radiation that strikes them (not more than 10 percent), so that most of what is not reflected is transmitted through them.

TABLE **3.1**   Reflectivity or "albedo" of various surfaces

Surface	Percent reflected
Cloud (stratus) < 500 ft thick	25–67
500–1000 ft thick	45–75
1000–2000 ft thick	59–84
Average of all types and thicknesses	50–55
Concrete	17–27
Crop, green	5–25
Forest, green	5–10
Meadow, green	5–25
Plowed field, moist	14–17
Road, blacktop	5–10
Sand, white	30–60
Snow, fresh-fallen	80–90
Snow, old	45–70
Soil, dark	5–15
Soil, light (or desert)	25–30
Water	8*

*Typical value for water surface, but the reflectivity increases
sharply from less than 5 percent when the sun's altitude above
the horizon is greater than 30° to more than 60 percent when the
altitude is less than 3°. Rough seas have a somewhat lower albedo than calm seas.

In general, the Earth's surface is a poor reflector of solar radiation, although the amount reflected varies greatly with the nature of the surface. Table 3.1 gives the reflectivity for a number of common surfaces.

Solar Energy Absorbed by the Earth

Figure 3.4 presents a summary of what happens, on the average, to the solar radiation intercepted by the earth. Of the 100 percent of the total solar radiation reaching the top of the earth's atmosphere (342 Watts m^{-2}), approximately 25 percent is absorbed by the dust, water vapor, and cloud droplets in the atmosphere. About 8 percent is reflected to space from the Earth's surface, which is a poor reflector. On the other hand, clouds and aerosols are relatively good reflectors, and, on the average, they reflect 24 percent of the sun's radiation back to space. Because clouds play such an important role in the global energy budget, variations in the average cloud cover are potentially important in producing climate changes.

The Earth's surface absorbs about 43 percent of the solar radiation: some of it comes directly from the sun, some after reflection by clouds, and the rest after being scattered by the air. The reflectivity of the Earth's surface varies greatly, of course. Some fresh snow fields and water surfaces (when the sun is close to the horizon) reflect 90 percent or more of the incident rays. But a forest may reflect less than 10 percent and green grass fields only 10–15 percent.

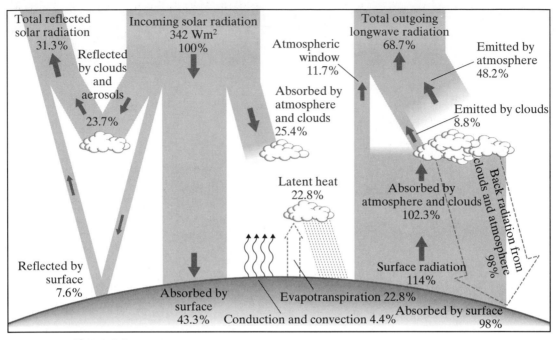

Figure 3.4 Earth's average heat.

Thus, of the total energy arriving from the sun (342 Watts m^{-2}), approximately 69 percent is absorbed by the Earth's surface and atmosphere. The rest, 31 percent, is lost to space, having been reflected by clouds and the Earth's surface or scattered by the particles in the air. The average reflectivity, "whiteness," or **albedo** of the Earth is said, therefore, to be 0.31, since that is the fractional part of the incident radiation that is bounced off the Earth. In contrast, the moon's albedo is only about 7 percent, which means that it is not nearly as bright as the Earth.

3.3   THE EARTH'S HEAT BALANCE

Studies indicate that over moderately long periods of time (between hundreds and thousands of years) the mean temperature of the earth is essentially constant. This indicates that there exists a long-term energy balance between the earth and space. It follows, then, that since 69 percent of the solar energy striking the earth is absorbed, an equal amount must be reradiated to space. Figure 3.4 shows what happens to the Earth's energy. Note that 69 percent is lost by the Earth and its atmosphere to space. Keep in mind, though, that although there is a heat balance for the planet as a whole, all parts of the Earth and its atmosphere are not in radiative balance. In fact, it is the imbalance between incoming and outgoing energy over the Earth that leads to the creation of wind systems and ocean currents that act to alleviate the surpluses and deficits of heat that would otherwise result.

Of special interest in the long-wave energy transfers is the fact that the amounts emitted by the Earth and the air actually exceed the total solar energy amount retained by the Earth (69 percent). This can be explained in terms of the "blanketing" or "greenhouse" effect of the atmosphere, which keeps the earth's surface and lower layers of the atmosphere a good deal warmer (35°C) than they would be without the atmosphere. For example, the moon's sunny surface, which absorbs almost twice as much energy per unit area as does the Earth's surface, is more than 20°C colder because it lacks an atmospheric "blanket."

Two gases—water vapor and carbon dioxide—play the most important role in keeping the Earth warm. Except for the "windows" between about 8 and 13 μm (see Figure 3.3), these gases block the direct escape of the infrared energy emitted by the Earth's surface. Although the atmosphere and clouds absorb only a small percentage of the short-wave solar radiation (25 percent, see Figure 3.4), it is quite opaque to the long-wave terrestrial radiation (absorbing 102/114 = 90 percent, see Figure 3.4). Only when the Earth's surface temperature is fairly high does the radiational loss through the transparent bands and from the top of the atmosphere equal the amount absorbed from the sun. This heat-retaining behavior of the atmosphere is somewhat analogous to what happens in a greenhouse. In a greenhouse, the glass (or more recently, plastic) roof and sides permit the sun's energy to enter and be absorbed by plants and Earth, but they prevent much of the interior heat from escaping by blocking mixing of the inside air with that on the outside. To a lesser extent, the glass also prevents escape of the energy radiated at long wavelengths from inside the greenhouse. The role of moisture, carbon dioxide and other trace gases in the atmosphere, allowing solar energy to pass through while absorbing much of the Earth's long-wave radiation, has thus come to be known as the **greenhouse effect**. This effect is quite noticeable when one compares the rapid temperature fall at night in the desert, where the air is dry, with the slower decrease in temperature in coastal regions, where the air is moist. On a global basis, the effectiveness of the atmospheric "greenhouse" is illustrated by the fact that the mean air temperature near the surface is about 13°C, while the overall "planetary temperature" is only about –20°C. Venus's greenhouse effect is even more dramatic. Its radiative temperature is about –30°C, but its surface temperature is estimated to be more than 450°C.

3.4 SEASONAL AND LATITUDINAL VARIATION OF THE EARTH'S HEAT ENERGY

The variation of the Earth's energy budget with latitude and with the seasons is a result of four factors: (1) the Earth is essentially a sphere; (2) the Earth rotates around the sun; (3) the Earth rotates about an axis connecting the North and South poles; and (4) the Earth is so far away from the sun that the sun's rays of light are approximately parallel.

The seasons are caused primarily by the motion of the Earth around the sun and the tilt of the Earth's axis of rotation as shown in Figure 3.5. The Earth's orbit around the sun is elliptical as shown in the figure, however, the degree of ellipticity is greatly exaggerated in this figure; the actual orbit is nearly a circle. The sun is at one of the foci of the ellipse, so both the maximum and minimum distances between the Earth

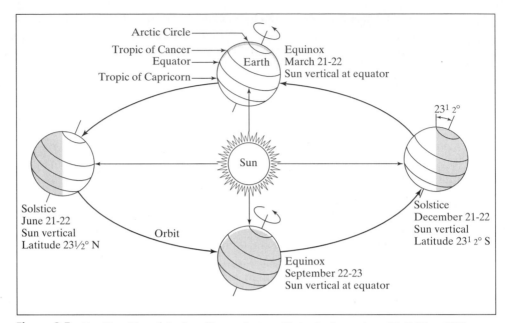

Figure 3.5   Earth's orbit and the Sun. (Source: Lutgens/Tarbuck, *Atmosphere*, 6th Edition, 1995, Prentice Hall.)

and the sun occur when the Earth is located on the major axis. Minimum distance, perihelion, occurs around January 5; maximum distance, aphelion, occurs six months later around July 5. The obvious result of this difference is that July should be colder than January. Yet, it is not—at least in the Northern Hemisphere. So the changing distance between the Earth and the sun cannot be the main reason for seasonal changes. The difference of about 3.5 percent in Earth-sun distance between aphelion and perihelion produces a difference in the total amount of solar energy received by the Earth of about 7 percent between January and July. The effect of this difference is small compared to the effect of the tilt of the Earth's axis of rotation with respect to the plane of its orbit around the sun. At present, this angle of tilt is approximately 23.5° (Figure 3.5). This angle, which is called the obliquity of the ecliptic, has varied over the past 100,000 years between 22 and 25° as the Earth rocks back and forth a little. The variation over many years of this angle causes important changes in climate.

As the Earth revolves around the sun, its axis points in the same direction in space. Figure 3.5 shows the Earth in four positions, the summer solstice, the autumnal equinox, the winter solstice and the vernal equinox (Note: these terms refer to the solstices and equinoxes in the Northern Hemisphere; the seasons are reversed in the Southern Hemisphere). On June 21 or 22, the Northern Hemisphere is directed toward the sun; the North Pole is in sunlight all day and the South Pole never sees the sun during the day.

The satellite photograph of the Western Hemisphere at 5:45 A.M. EDT on June 21, 1995 (Figure 3.6) illustrates the effect of the tilt of the Earth's axis on the length of night and day. Because this date is the summer solstice, the North Pole is bathed

CSU/CIRA

June 21, 1995
1145Z

Figure 3.6 Visible satellite (GOES-8) photograph for 5:45 E.D.T., June 21, 1995. (Source: The Cooperative Institute for Research in the Atmosphere at Colorado State University.)

in sunlight, and the South Pole is in darkness. The satellite is centered over the equator at 45°W longitude, so the line of sunrise makes an angle of nearly 23.5° with the
45° meridian. Thus the comma-shaped mass of swirling clouds over the northern Atlantic with a trailing cold front that connects a massive cloud system off the east coast
of the United States is illuminated by the early morning sun. Farther west, central
and eastern Canada, the upper midwest of the United States and New England are
bathed in daylight, while locations at the same longitude but farther to the south,
such as New Orleans, have an hour or two of night left.

The major seasonal changes we now observe would not exist if the obliquity of
the ecliptic were zero. Over tens of thousands of years this angle has changed, and as
a result the severity of the seasons has also changed. When the angle is small, the seasons are less extreme than when the angle is large.

As mentioned above, although the overall income and outgo of radiant energy
are essentially in balance, they are not in balance everywhere on the Earth. This is
mainly because the amount of incoming energy varies greatly from place to place. It
is also caused, to a lesser extent, by variations in the intensity of outgoing radiation.
The amount of energy *emitted* by the Earth and the atmosphere to space is controlled
largely by the amount of moisture in the air; the distribution of solar energy *absorbed*
by the Earth's surface is controlled mostly by the Earth's movements, the distribution
of physical properties of the surface, and cloudiness.

The latitudinal variations of absorbed solar energy is illustrated in Figure 3.7.
Only one-half of the sphere can be illuminated at one time, and the angle that the sun's
rays make with the sphere's surface will decrease from 90 degrees at the exact center of the lit hemisphere to 0 degrees at the edges (where the shadow begins). This is
illustrated in Figure 3.7(a). Angle c equals 90 degrees at one point where the sun's rays
are perpendicular to the surface, while angle b is less than 90 degrees, and angle a is
less than angle b. The rays entering with the angle a will, of course, traverse a greater
mass of atmosphere than those entering with angle b or angle c (since distance $d_1 >
d_2 > d_3$), and therefore they will be subject to greater depletion by absorption, re-
flection, and scattering. But even more important, the intensity will be less at lower
solar angles because the same amount of energy will intercept larger surface areas,
as is illustrated in Figure 3.7(b), so that the amount of energy received by each square
kilometer of surface will be less for low angles of the sun than for high ones.

Figure 3.8 shows the way in which the total solar energy (**insolation**) received
each day varies with time of year and latitude. Note that around the time of the summer **solstice** in each hemisphere, the total daily energy received varies little between
the poles and the equator: this is because the lower solar angles in the polar regions
are compensated for by the greater duration of sunshine each day. In the winter, of
course, the latitudinal variation in the amount of energy received is very large, since
practically none is received at high latitudes while the equatorial region's supply remains almost unchanged throughout the year.

The amount of incoming and outgoing energy, averaged over the entire year, is
shown for each latitude in Figure 3.9. It can be seen that, as would be expected, the
incoming short-wave energy decreases a great deal between the equator and the
poles, while the outgoing long-wave energy is nearly constant. Averaged over all latitudes, the incoming energy (curve I) equals the outgoing energy (curve II). But there

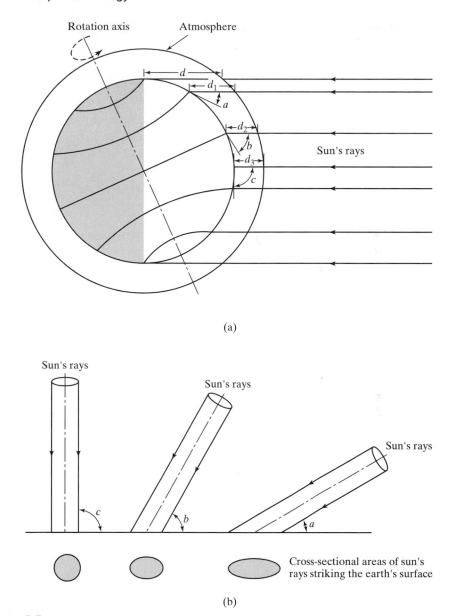

(a)

(b)

Figure 3.7 The intensity of solar radiation depends on the angle at which the Sun's rays strike Earth's surface. (a) The angles of incidence a, b, c and the depths of penetration through the atmosphere, d, d_2, d_3, at different parallels of latitude. (b) The varying cross-sectional area on Earth's surface due to different angles of incidence.

is a surplus of energy income at tropical latitudes and a deficit in the polar regions. If air movements (and, to a lesser extent, ocean currents) did not exist to redistribute the energy, the poles would become steadily colder and the tropics steadily warmer, until a new radiative equilibrium was established.

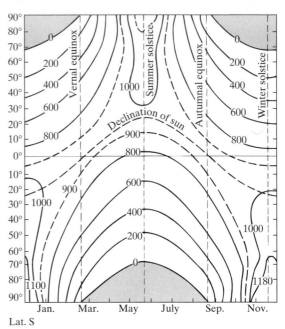

Figure 3.8 Undepleted insolation in cal/cm²/day as a function of latitude and date. Shaded areas represent latitudes entirely withinEarth's shadow.

If there were only latitudinal variations in the energy received and absorbed by the Earth and its atmosphere, meteorology would be a considerably simpler study. However, the absorptive properties of the air and the surface of the Earth are not distributed in a smooth, unchanging pattern. The radiation absorption properties of the atmosphere depend on the clouds, moisture, and dust in the air, the concentrations of which may vary both geographically and with time. Surface properties are also erratically distributed over the face of the Earth, and even they change with time.

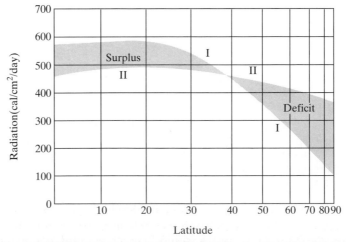

Figure 3.9 Curves I and II represent mean annual insolation and outgoing long-wave flux, respectively, at the tropopause.

The most pronounced differences in surface thermal properties are those between land and sea. Under identical insolation conditions (same solar angles, duration of daylight, atmospheric transparency), the temperature changes experienced by the water are much less than those of the land surface. The principal reason for this is that water is a fluid and so can be mixed; as a result, its heat tends to be distributed over a much greater mass than is the case with solid land. The heat absorbed by a land surface tends to be confined in the upper few centimeters, while in water the heat may be distributed to depths of hundreds of meters. There are other reasons for the smaller temperature range of water surfaces: (1) because water is transparent, radiation can penetrate to depths of tens or even hundreds of meters, so that the energy is absorbed by a great mass of water, (2) water has a higher specific heat than does land (i.e., more heat is required to raise the temperature of a gram of water $1°C$ than for land), and (3) some heat is used in evaporation of water (latent heat).

The fact that the oceans act as heat reservoirs is illustrated by the January and July mean air temperature maps of Figure 3.10. Note how much more the temperature varies between seasons in middle and high latitudes over the continents than it does over the oceans. For example, at latitude 45°N, the annual range of temperature over the continents is about 60°F, while over the Pacific Ocean at the same latitude, the range is only about 10°F. Note also how the isotherms dip equatorward over the oceans in summer and poleward in the winter, indicating that the ocean is cooler than the land in summer and warmer than the land in winter.

Temperature Lag

The times of high and low air temperatures do not coincide with the times of maximum and minimum solar radiation, either on an annual or a daily basis. The months of July and August are generally the hottest of the year, while January and February are the coldest; yet the greatest intensity of radiation occurs in June and the lowest in December. On a daily basis, the highest temperature normally occurs around 2 P.M., yet the greatest intensity of insolation each day occurs near noon.

This lag in temperature is explained by Figure 3.11. The Earth loses heat continuously through radiation. During some months of the year and some hours of the day, the incoming energy exceeds the outgoing energy of the Earth. While this is occurring, the temperature will increase because the air's heat content will be rising. The maximum temperature will occur at the time when the incoming energy ceases to exceed the outgoing. Thereafter, when the outgoing energy is greater than the incoming, the temperature will fall until the two are again in balance. At the point where a "surplus" of energy begins to appear, the lowest temperature will have occurred.

3.5   THE POSSIBLE USE OF SOLAR ENERGY

Concern over the increased cost and limited supply of oil and other fossil fuels as energy sources has intensified interest in the possibilities of the direct use of solar energy. Both the potential and some of the limiting factors are evident from the discussion in the previous sections of this chapter: An average of about 0.5 cal/cm^2/min is inter-

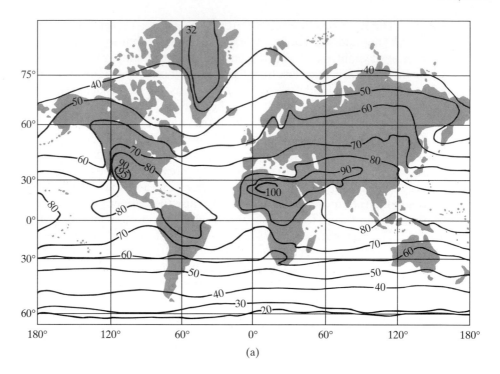

(a)

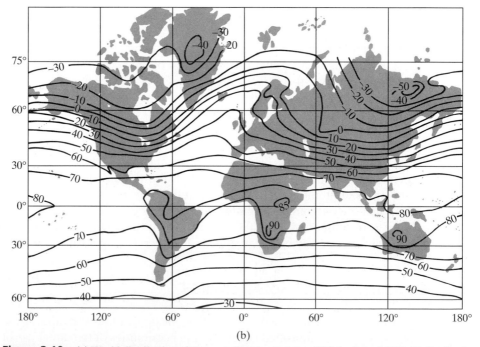

(b)

Figure 3.10   (a) World distribution of mean surface temperature (°F) for July; (b) World distribution of mean temperature (°F) for January.

The Temperature Lag

The Daily Cycle

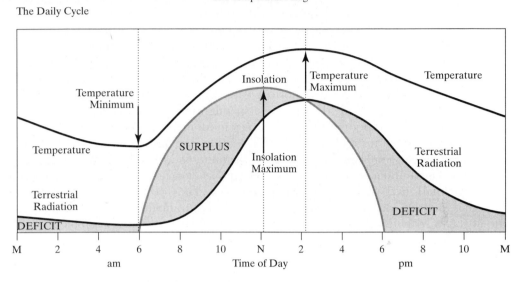

The Annual Cycle

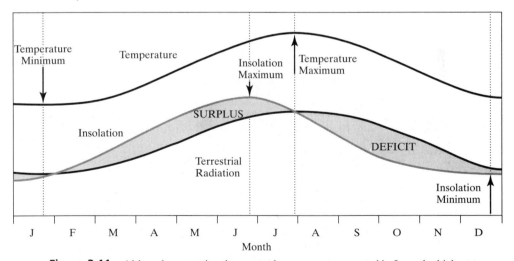

Figure 3.11   Although we receive the most solar energy at noon, and in June, the highest temperatures do not occur until a later time. The reason for this delay is that temperature depends upon the balance between incoming energy and energy lost. Even after the time of maximum insolation, we
continue to receive more energy than the ground loses, so the temperature continues to rise during this
"temperature lag." After the energy received and energy lost reach equilibrium, energy lost becomes
greater, so the temperature starts to decrease. (Courtesy Thomas McGuire.)

cepted by the entire Earth (see Problem 1 at end of chapter); of this amount, 50 percent (see Figure 3.4) or 0.25 cal/cm^2/min, reaches the surface of the Earth. This comes
to about 171 watts on every square meter of the Earth's surface (see Appendix 1 for

conversion of the units), or more than 4 kilowatt-hours per day. This amount of energy is about half as much as is required to heat and cool the average U.S. house. In other terms, if the 4 kilowatt-hours per square meter could be converted to electricity with an efficiency of 10 percent, some 400,000 kilowatt-hours of electricity could be obtained from a solar collection area of 1 square kilometer; this represents about $\frac{1}{75}$ (1.33 percent) of the energy consumed by a city with a population of from 1 to $1\frac{1}{2}$ million. All the U.S. electrical energy could be supplied with solar energy collectors covering 0.14 percent of the land, again assuming 10-percent efficiency.

Although these computations are correct, they are based on the *average* incidence of solar energy on the Earth's surface. The amount of sunshine intercepted varies greatly with latitude and season (see Figure 3.8) and time of day; the amount reaching the surface depends, of course, on the atmospheric transparency—primarily cloudiness—which varies greatly from day to day. Thus, one of the greatest problems in using solar energy is its intermittency. Complete reliance on solar energy requires the development of techniques for storing energy. This will probably involve the creation of synthetic fuels from solar energy.

The heating of buildings and water with solar energy is already being done and is relatively simple. A dark radiation-absorbing surface is exposed to the sun's rays, perhaps mounted on sloping roofs; panes of glass are mounted above the surface to trap the heat. Air or water pumped over, under, or through the heated surface transports the heat to the interior of the building or to water tanks.

There are many other "indirect" solar energy sources; examples are the conversion of plant material to fuels, the harnessing of ocean tides and waves, and the utilization of winds. Each of these potential sources contains only a very small proportion of the total solar power reaching the Earth's surface. A sample calculation of the energy extractable from the wind is presented in Problem 11, at the end of this chapter.

KEY TERMS

absorption
albedo
aphelion
black body radiation
conduction
convection
electromagnetic spectrum

greenhouse effect
heat energy
insolation
kinetic energy
long-wave radiation
perihelion
radiation

reflection
scattering
short-wave radiation
solstice
Stefan-Boltzmann law
Wien's law

PROBLEMS

1. Compute the total energy per minute intercepted by the Earth and the fraction of the total solar energy output this represents. (Hint: The cross-sectional area of the Earth = πr^2 = $3.1416 \times (6000)^2 km^2$, while the surface of an imaginary sphere surrounding the sun at the distance d of the Earth from the sun = $4\pi d^2 = 4 \times 3.1416 \times (149.6 \times 10^6)^2 km^2$. (The solar constant = 1370 W/m^2.)

2. If you were attempting to observe the temperature distribution on the moon's surface by measuring the infrared radiation, what wavelength band would give the best results, considering the atmosphere's transparency?

3. Compute the elevation angle of the sun at noon at the latitude of your city on Dec. 21. What will be the length of the daylight (not counting twilight)?

4. Explain why nighttime temperatures are generally lower on nights when the humidity is low than when it is high.

5. What is the essential difference between the phrases "light from the sun" and "radiation from the sun?"

6. What angle between the Earth's axis of rotation and the plane containing the Earth's path around the sun would provide (a) the least difference between seasons? (b) the greatest difference?

7. Suppose you wish to design your living room so that it receives as much sunshine as possible during the winter, and as little as possible during the summer. If you live in the Northern Hemisphere, in which direction should the windows face? What if you live in the Southern Hemisphere?

8. How do you explain the difference in albedo between the earth and the moon? Is the dark side of the moon illuminated by "earthlight," as we are by moonlight?

9. How much heat per unit area is required to melt a layer of snow at a rate of 1 in/day, assuming a liquid water equivalent of 10:1? What percentage of the mean total outgoing radiation at latitude 50 degrees does this represent?

10. The efficiency of an engine is defined as the ratio of the rate at which work is extracted to the rate at which energy is added. Use the results of Problem 1 and the fact from Figure 3.4 that 69 percent of the incident solar radiation is absorbed by the earth and its atmosphere to show that the average rate of energy addition to the Earth-atmosphere system is 244 W/m^2. If the rate at which kinetic energy is generated in the atmosphere is about 2 W/m^2, how efficient is the atmospheric heat engine? How does this compare with the average automobile engine?

11. Wind power has long been used to do a very minor portion of people's work. This is not too difficult to understand. Suppose we have a windmill that always faces into the wind and has a radius of 5 meters (area = 78.5 m^2). It drives an electric generator. Assuming an average wind speed of 5 m/s (11.2 mi/h), the air's energy per gram is $\frac{1}{2}v^2 = 12.5$ m^2/s^2. The volume of air intercepted by our windmill in one year (3.1536×10^7s) equals the distance traveled by the air times the windmill area = (5 m/s) $\times$ (3.1536×10^7s) $\times$ (78.5 m^2) = 1.2378×10^{10} m^3. Since the average air density is 1.2 kg/m^3, the number of kilograms striking the blades is ($1.2378 \times 10^{10}m^3$) $\times$ 1.2 kg/m^3) = 1.485×10^{10} kg. Thus the total possible energy is (1.485×10^{10} kg) $\times$ 12.5 m^2/s^2 or 18.56×10^{10} kg m^2/s^2 or 18.56×10^{10} joules. Since 1 joule equals 2.777×10^{-7} kWh, the total possible energy is 5155.5 kWh. If the efficiency of energy capture and conversion to electricity were as high as 10 percent, the useful energy generated in one year would be 516 kWh. How does this compare with the average household consumption? Is it practical? What other factors (for example, characteristics of the wind) have not been considered?

12. The axis labeled "latitude" in Figure 3.9 is scaled as the cosine of latitude rather than linearly with latitude. Why?

4

Air in Motion

The uneven distribution of heat resulting from latitudinal variations in insolation and from differences in absorptivity of the Earth's surface leads to air motions. The mechanics of this conversion of heat energy to kinetic energy will be the subject of this chapter.

We will be examining the deviations of air motion from those that are due to the planetary motion. The atmosphere as a whole follows the Earth in its movements through space; it also rotates with the Earth from west to east, so that at the equator the air moves eastward at a speed of more than 1600 km/h, while at latitude 60 degrees it moves eastward at half that speed. Of course, at the poles its eastward speed is zero. Because the ground moves at the same eastward speed, these motions go unnoticed by the Earthbound observer. Winds are those motions of the air *relative* to the Earth.

4.1 PRINCIPAL FORCES IN THE ATMOSPHERE

According to **Newton's first law** (law of inertia), for a body to change its state of motion, it must be acted upon by an unbalanced force. According to **Newton's second law**, the force required to accelerate a body of mass m is given by

$$F = ma \tag{4.1}$$

where a is the acceleration (rate of change of velocity). There are two classes of forces that affect the atmosphere: (1) those that exist regardless of the state of motion of the air and (2) those that arise only *after* there is motion. The first category can be thought of a the fundamental of basic forces, since without them there would be no motion. These basic "driving" forces are produced by gravitational attraction and pressure. In the second group of forces are friction or "drag" and Coriolis forces.

4.2 THE DRIVING FORCES—GRAVITY AND PRESSURE

In order to understand the forces that cause air to speed up or slow down, and the balance of forces that is associated with steady currents of air, it is convenient to consider the forces on a small "parcel" of air, for example a small piece or chunk of air weighing one gram. The net of all the forces acting on this parcel of air determines its acceleration and its motion. Let us consider forces acting in the vertical direction on this parcel of air.

Forces Acting in the Vertical Direction

The **gravitational** pull of the Earth, or the gravitational force, F_g, is always directed downward, toward the center of the Earth

$$F_g = Mg \qquad (4.2)$$

where M is the mass of the parcel of air and g is the acceleration of gravity (9.8 m s^{-2}). Gravity produces a downward acceleration on parcels of air (Figure 4.1), and if it were not balanced by an upward directed force, the parcel of air would be accelerated rapidly downward. Since on the average the air is not moving upward or downward, there must be another force directed upward that tends to balance gravity. This force is the vertical component of the **pressure gradient force** and is associated with the large decrease of air pressure with height.

When a difference in pressure exists in a fluid such as air or water, a force is exerted from high to low pressure. The magnitude of this force is directly proportional to the difference, or gradient, in the pressure. In the atmosphere, the pressure always decreases with height (since the mass of the air remaining above decreases with height). In Figure 4.1, the pressure P_1 on the bottom side of the parcel is greater than the pressure P_2 on the top side of the parcel, which results in an *upward* acceleration. This upward directed force associated with the decrease of pressure with height is

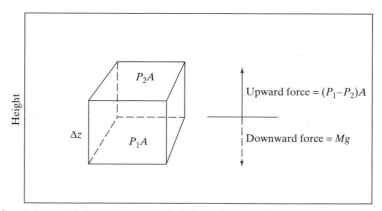

Figure 4.1 Hydrostatic balance represents the balance between the vertical component of the pressure-gradient force and gravity.

always nearly balanced by *downward* force of gravity; it is the small difference between these two oppositely-directed forces that determines whether a parcel of air will
move upward or downward, and at what rate.

The net pressure gradient force in the vertical is given by the difference in pressure $P_1 - P_2$ times the surface area of the top and bottom faces, A. A good approximation under most conditions shows that the vertical pressure gradient force and
gravity are in balance, a condition called **hydrostatic balance**. This balance is represented by equating the two forces.

$$(P_1 - P_2) \times A = Mg \qquad \textbf{(4.3)}$$

Since the mass M equals the density ρ times the volume $A\Delta z$, (4.3) can be written

$$(P_1 - P_2) / \Delta z = \rho g \qquad \textbf{(4.4)}$$

which is the **hydrostatic equation**.

Pressure Gradient Force in the Horizontal Direction

Although the most rapid variation of pressure with distance occurs in the vertical direction, pressure also varies in the horizontal direction and these relatively small horizontal variations are extremely important in producing atmospheric motions because
they are *not* balanced by gravity. Figure 4.2 illustrates a typical pressure pattern in a
vertical cut of the atmosphere. In the figure the vertical component of the pressure
gradient force, obtained by dividing the difference in pressure between points C and
A by the vertical distance between these points, is 24 mb/200 m, or 120 mb/km. The
horizontal component, obtained by dividing the difference in pressure between points

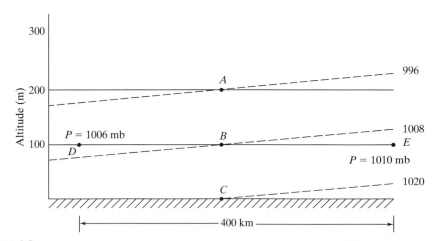

Figure 4.2 Horizontal and vertical variation of pressure. Dashed lines are isobaric surfaces labeled
in millibars (mb).

E and *D* by the horizontal distance between these points, is 4 mb/400 km, or 0.01 mb/km. Thus the vertical component of the pressure-gradient force is 12,000 times greater than the horizontal component in this typical example.

In order to represent horizontal pressure gradients on a weather map, the pressure must be depicted on a surface of constant height. The most common weather map is the sea-level pressure map, in which lines of constant pressure (isobars) are plotted on a map representing sea level. Since most points on land are not at sea level, it is necessary to calculate the pressure at each weather station that would exist if the elevation of that station were reduced to sea level. This process is called **reducing** the actual surface pressure to sea-level pressure.

Figure 6.1 shows a sea-level pressure map for 7:00 E.S.T. on March 13, 1993, the day of a major blizzard on the East Coast of the United States (this storm will be discussed in detail in Chapter 6). The isobars in Figure 6.1 are plotted at 4 mb intervals. The horizontal spacing between the isobars is inversely proportional to the horizontal pressure gradient force; where the isobars are close together, such as in northern Georgia and Alabama, the pressure gradient is large. Where the isobars are widely spaced, as in northern Minnesota, the horizontal pressure gradient is small.

The direction of the horizontal pressure gradient force is always directed from high to low, hence it is oriented at right angles to the isobars. Several vectors (arrows) representing the horizontal pressure gradient force are shown in Figure 6.1.

The horizontal pressure gradient in the atmosphere usually varies considerably in the vertical. In middle latitudes during most seasons it generally increases in magnitude with height up to a level of 8 km or so. The reason the horizontal pressure gradient varies with height is because there are usually horizontal differences, or gradients, in temperature. This relationship may be seen from the hydrostatic equation, and is illustrated in Figure 4.3.

The hydrostatic equation states the variation of pressure in the vertical is greater in cold air (high density) than in warm air (low density). In Figure 4.3, for example, the isobaric surfaces are closer together to the north where the density is high. Therefore, the high pressure (on a constant height surface) that exists near the surface in

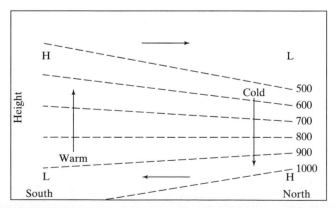

Figure 4.3 Variations in the separation of isobaric surfaces between warm and cold air lead to reversal of pressure systems with height. Dashed lines in isobars (mb). Direction of circulation is indicated by arrows.

the south becomes a high aloft. This principle can be seen in the climatological distribution of pressure; high pressure at the surface over polar regions is accompanied by low pressure aloft, while at the equator low pressure at the surface is overlain by high pressure aloft. On a smaller scale, cold anticyclones (highs) at the surface weaken with height and become cyclones (lows) aloft; while warm cyclones (such as hurricanes) at the surface weaken and become anticyclones aloft.

In the example shown in Figure 4.3, the warm, less dense air in the south will tend to rise while the cold, more dense air to the north will tend to sink. Near the surface, the horizontal pressure gradient is directed from north to south and, in the absence of other forces, the air will tend to be accelerated toward the south. However, the northward slope of the isobars decreases with altitude above the surface because the air is colder and denser toward the north and the isobars are packed more closely in the vertical. Eventually, the northward slope is reversed and at high altitudes the horizontal pressure gradient is directed from south to north. This is because the vertical pressure gradient in cold air is greater than it is in warm air, as shown by the hydrostatic equation (4.4). At any particular pressure, the density is inversely proportional to temperature [equation of state (1.4)]; warm air is less dense than cold air.*

The simplified closed circuit of air motion represented in Figure 4.3 is a *thermal circulation*. It represents the circulation that would be generated on a simple, nonrotating planet which was cooled at the North Pole and warmed at the Equator. Note that resembles the pattern of motion that occurs in a tank of water heated on one side.

In summary, there are two rather large 'driving' forces acting on the atmosphere—pressure gradient and gravity. The latter is directed entirely in the vertical, while the former has a very small component directed in the horizontal. Even though each of the two forces acting in the *vertical* is much greater than in the *horizontal*, this does not mean that vertical motion is much stronger than horizontal motion; it is *net* or unbalanced force that determines acceleration, and the two vertical forces are almost always very nearly equal and oppositely directed. Actually, except in relatively small circulation cells such as those of a thunderstorm, the vertical air velocity is normally only a tenth or a hundredth of the horizontal velocity.

4.3   FORCES THAT ARISE AFTER THERE IS MOTION

The Effect of the Earth's Rotation

Large-scale flow in the Earth's atmosphere does not follow the simple pattern of the thermal circulation shown in Figure 4.3. The horizontal wind blows more nearly *perpendicular* to the pressure gradient than along it. Figures 4.4 and 4.5 show upper-level (500 mb) weather maps for the Northern and Southern Hemispheres, with arrows designating the observed wind direction and speed (see Appendix 4 for plotting model). Note that the air is moving more or less *along* the height contours, rather than across them. (Height contours on a constant pressure sm force such as 500 mb

---

*The density also depends slightly on the moisture in the air. A mixture of air and water vapor is less dense than dry air because water has a lower molecular weight—18—than the average for dry air—29.

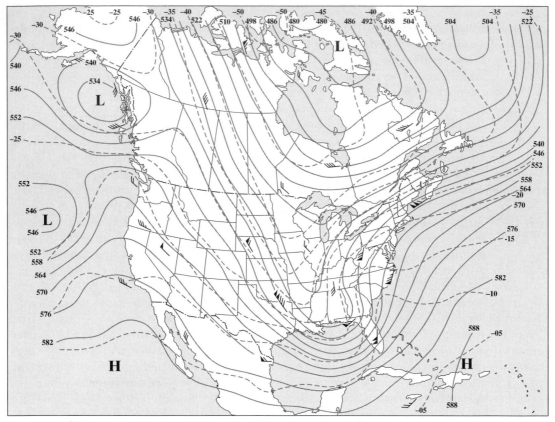

Figure 4.4 500-mb map for 7:00 A.M., E.S.T., March 13, 1993.

are analogous to isobars on a constant height surface.) Evidently something steers the air flow to the right of its target (center of low pressure) in the Northern Hemisphere and to the left in the Southern Hemisphere. The only thing that could effect this difference between the two hemispheres is the Earth's rotation. This is illustrated by the turntable of Figure 4.6. The top of the turntable has been given the same sense of rotation (counterclockwise) as the Earth's Northern Hemisphere; however, if one looks at the same turntable from below, the sense of rotation is opposite (clockwise).

Imagine yourself on a very large merry-go-round whose rate and sense of rotation can be varied. This is the situation of an Earthbound observer (Figure 4.7). At the North Pole, an observer is spinning through the vertical axis at a rate of one rotation per day in a counterclockwise sense. At the South Pole, an observer is also spinning through the vertical axis at the rate of one rotation per day, but in the opposite (clockwise) sense. At the equator, the observer is not spinning at all around his vertical axis, although his "merry-go-round" is turning end over end at the rate of one complete turn per day. At some latitude intermediate between the pole and the equator, the rate of rotation around the vertical axis is something between the one rotation per day at the pole and the zero rotation per day at the equator. It is the ro-

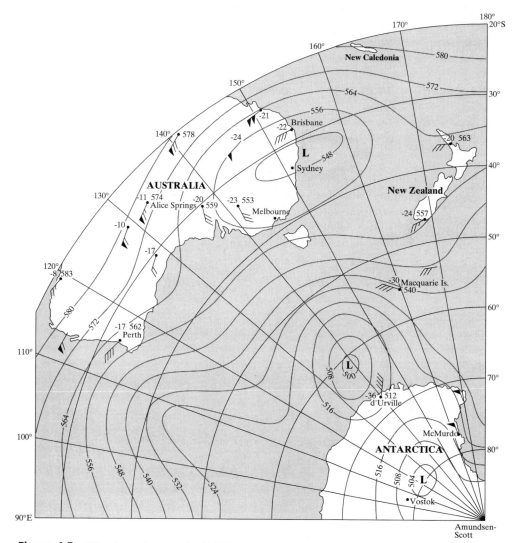

Figure 4.5   500-mb weather map for 1200 UTC, June 30, 1958. Wind speed and direction at selected stations indicated by arrow shafts and barbs (full barb = 10 knots). Temperatures plotted in °Celsius. See Appendix 4 for data plotting convention. Contours are heights of the 500-mb surface in decameters.

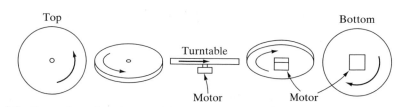

Figure 4.6   Sense of rotation of a turntable as seen from above and below.

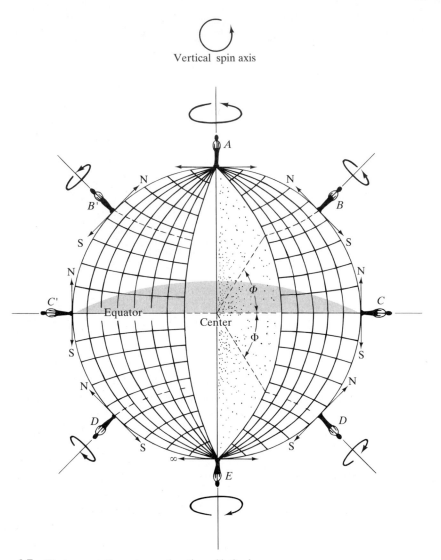

Figure 4.7   Horizon rotation rate as a function of latitude.

tation around the vertical axis which concerns us most, because it is what affects the horizontal motion and causes air to deviate from a high-to-low pressure path. If the latitude is designated by ϕ and the Earth's rate of rotation (1 rotation/day) is designated as ω, the rate of rotation around the vertical axis can be shown to be $\omega \sin \phi$.

   Figure 4.8 shows what happens if you try to throw a ball from a rotating merry-go-round at some target riding near the edge of the platform. To the observer on the merry-go-round, it will appear that some force has caused the ball to curve to the right of the intended path in the case of counterclockwise rotation (to the left for clockwise rotation). An observer not on the merry-go-round would say that the ball

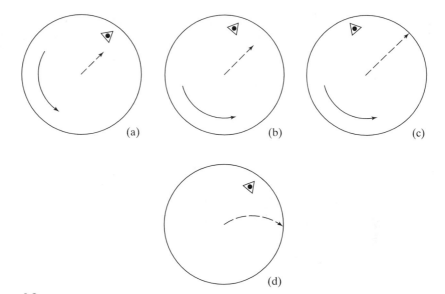

Figure 4.8 Effect of rotation on apparent path of a moving body. A projectile is fired from the center of a merry-go-round at a triangular target (a). As the target rotates (b,c), the projectile misses the target to the right. To an observer rotating on the merry-go-round, the projectile appears to curve to the right (d).

moved along a straight line but that the target turned. The observer on the platform could account for the deflection of the ball from the target by supposing a deviating force. The same thing happens on the rotating Earth—the line connecting the target (a low-pressure area) and a parcel of air is continuously changing its orientation.

The fictitious deviating force (fictitious *only* as far as a nonrotating observer is concerned) introduced to account for the effect of the Earth's rotation is known as the **Coriolis force**, named for the French mathematician who first explained it mathematically. The algebraic expression for the apparent acceleration is $2V\omega \sin \phi$, where V is the speed of the particle relative to the Earth's surface. The magnitude of the acceleration is thus not only dependent on the latitude (maximum at the poles, zero at the equator), but also directly proportional to the speed of the particle. It always acts at a 90-degree angle to the wind. Although the Coriolis force is small, it is significant in horizontal air flow because, first of all, the horizontal pressure gradient force is also relatively feeble and, second, the air traverses great distances. It is very important in the large-scale flow, such as that associated with systems that affect the weather over thousands of miles, but it is of much lesser consequence in local, small-scale winds. The same, of course, is true in its effect on any body moving freely over the Earth's surface. The correction must be applied to long-range artillery, but it has no detectable effect on the path of a bullet. Ocean currents are appreciably affected.

The Coriolis force comes into effect as soon as a particle has motion. For all moving parcels of air away from the equator, the Coriolis force constantly accelerates the air to the right (in the Northern Hemisphere) or to the left (in the Southern Hemisphere). We can think of this apparent force as real, since we are concerned with the air motion relative to the Earth's surface.

At elevations more than a kilometer or so above the ground, frictional effects are small, and a near balance exists between the horizontal pressure gradient, directed toward low pressure on the left of the direction of motion, and the Coriolis force, directed toward the right of the motion (Figure 4.9). This important balance is called **geostrophic equilibrium**, and a wind in perfect balance with the horizontal pressure gradient force is called the **geostrophic wind**. Mathematically the geostrophic balance is given by

$$fV_g = \frac{1}{\rho}\frac{\Delta P}{\Delta n} \qquad \text{(4.5)}$$

where $f = 2\omega \sin \phi$ and $\Delta P/\Delta n$ is the magnitude of the horizontal pressure variation measured perpendicular to the flow. For example, the value of ω is $7.27 \times 10^{-5}\text{s}^{-1}$ so that at 43°N f is $1.0 \times 10^{-4}\text{s}^{-1}$. For the pressure gradient shown in Figure 4.9 of 4 mb/400 km $(1 \times 10^{-3}\text{ N/m}^3)$ and density ρ of 1.2 kg/m^3, the value for the geostrophic wind is

$$V_g = \frac{1.0 \times 10^{-3}\,\text{N/m}^3}{(1.0 \times 10^{-4}\text{s}^{-1})(1.2\text{ kg/m}^3)} = 8.3\text{ m/s} \qquad \text{(4.6)}$$

Curved Paths and Vortices

According to Newton's first law of motion, the velocity of a body does not change as long as there is no net (unbalanced) force acting on it. But keep in mind that velocity is a vector; i.e., it has both magnitude (speed) and direction. Thus, an object that

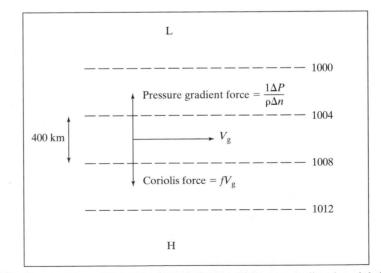

Figure 4.9 Geostrophic equilibrium in which the horizontal pressure gradient force is balanced by the Coriolis force.

moves at constant speed along a curved path is changing direction and is, therefore, experiencing an acceleration. This acceleration is directed at 90 ° to the path, inward along the turning radius. It is called **centripetal** acceleration.

A centripetal force must be applied to make an object deviate from its natural tendency to move along a straight line. For example, the muscles of a discus thrower apply the centripetal force needed to make the discus move on a circular path until the moment that he releases it; then, it moves along a straight line. We can think of this **centripetal force** that is making the object follow a curved path as opposed to a "**centrifugal** force" that is apparently pulling the object outward from the desired path.

Air parcels experience a centripetal acceleration whenever they travel along paths that are curved relative to the Earth's surface; they are forced to follow such a path whenever the isobars are curved or circular. The magnitude of this acceleration is given by V^2/r, where V is the wind speed and r is the radius of curvature of the air parcel's path. Thus, the tighter the curve and the faster the wind speed, the greater the centripetal force required. In air flow having gentle curvature, such as that in a circulation pattern that covers half a continent or more (see next chapter), the centripetal force is negligible compared to other forces. But in the case of a small, intense whirl, such as a tornado, in which r may be only 100 meters or less, the centripetal force can be very large.

The balance of forces in idealized frictionless flow around a circular low and high pressure system is shown in Figure 4.10. The wind that results from this balance of forces in curved or circular flow is called the **gradient wind**. In the case of the low, the pressure gradient force is directed inward toward the center of the low and to the left of the direction of motion. The Coriolis force is directed toward the right of the direction of motion. The magnitude of the pressure gradient force in this case exceeds that of the Coriolis force; the net difference is the **centripetal force** required to cause the air to turn and flow around the low parallel to the circular isobars.

In the case of flow around a circular high, the pressure gradient force is directed outward, away from the center of high pressure. The Coriolis force, directed to the right of the flow, now exceeds the pressure gradient force. Again, the difference is the centripetal force required to turn the air as it flows around the high.

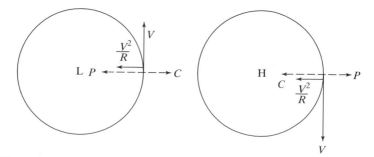

Figure 4.10   Balance of forces around cyclones and anticyclones. P is pressure gradient force, C is Coriolis force, and V^2/R is required centripetal force.

In the example shown in Figure 4.10, the pressure gradients associated with the high and the low are assumed to be equal. However, the Coriolis force is greater in the case of the flow around the high than for the flow around the low. The only way this can occur (at the same latitude) is for the wind speed to be greater around the high than the low. Therefore, *for the same horizontal pressure gradient and for the same latitude, winds are faster around highs than lows.*

The concept of the gradient wind applies to a variety of vortices in the atmosphere. The largest of these are synoptic scale cyclones and anticyclones which are major weather producers in middle latitudes. The diameter of cyclones and anticyclones is variable, but 2000—3000 km are typical values. Typically, the wind speed around cyclones and anticyclones near the surface do not exceed 70 km/h (45 mi/h). Smaller in size (average diameter about 700 kilometers) and much more destructive is the **tropical cyclone** or hurricane. The maximum surface wind speed in a hurricane is sometimes more than 200 km/h (125 mi/h). The smallest vortex, but the one with the most powerful punch, is the **tornado**. The intense rotation of this vortex is confined normally to a diameter of a kilometer or less, but its wind speed can reach 300 km/h (200 mi/h).

Because extratropical and tropical cyclones are large enough to be affected by the Earth's rotation, their winds are always cyclonic* about the center of low pressure. Small tornadoes (and their wet cousins, waterspouts) are too small to be affected by Coriolis forces; anticyclonic winds have been observed in these storms.

Because the centripetal force is produced by the pressure-gradient force for circulations about lows (see Figure 4.10), and there is no theoretical limit to the magnitude of the pressure gradient force, the speed of winds around lows is nearly unlimited. Around highs, however, the balance of forces is such that the centripetal force must be produced by an excess of the Coriolis force over the pressure gradient force. Thus balanced flow around highs can occur only on large scales. Furthermore, since the required centripetal force for a given radius increases as the square of the wind speed, while the Coriolis force only increases as the first power of the wind speed, balanced flow around highs can occur only at relatively low wind speeds.

Friction

Everyone is familiar with the fact that if a wooden box is given a push along a level floor, it will travel a short distance and then stop. The force that retards the forward motion is **friction**. It is the result of the interlocking of surface irregularities and the adhesion of touching molecules of the two contacting surfaces.

Although the adhesion is much less and the space between molecules is much greater in a gas, there is, nevertheless, a frictional drag created when velocity differences arise within a gas. This retardation of motion in a fluid is referred to as **viscosity**. When only the random, thermal motion of the *molecules* is responsible for this slowing up, the retardation, sometimes called **molecular viscosity**, is quite low. The ef-

---

*A cyclonic circulation is one in which the rotation is the same as that of the local horizon about the vertical. Thus cyclones rotate counterclockwise over the Northern Hemisphere and clockwise over the Southern Hemisphere.

fect of molecular agitation can be explained as follows: If a stream of air is directed along a solid surface, the air molecules in contact with the surface will have no motion (other than the usual random agitation) because they are blocked by the stationary molecules of the surface. These molecules will, in turn, retard the flow of those molecules adjacent to them because there is a continuous exchange of zero-velocity "surface" molecules with those in the next tier. Some of the slow-moving molecules of the second tier mix with those of the third tier, and so on, causing a progressively lesser retardation with distance from the surface.

The molecular viscosity of air is so small that if it alone were responsible for frictional drag in the atmosphere, the slowing up of the air flow would almost completely disappear within a meter of the surface. Far more significant is the so-called **eddy viscosity** which, at least in the lower layers of the atmosphere, is about 10,000 times more effective than molecular viscosity. As the name implies, it acts through the transfer of momentum between layers of air by **eddies** rather than by molecules.

Eddies are parcels of air that leave their places within otherwise orderly, smooth-path flow. A wind record marks the passage of these eddies as rapid irregular fluctuations in direction and speed. Figure 1.10 shows these fluctuation as recorded by an anemograph. Such fluctuations—deviations from the mean velocity—are referred to as **turbulence**. In addition to affecting momentum, turbulent fluctuations greatly accelerate the transport of the other components of a fluid. The most visible of these are pollutants, such as smoke. As a puff of smoke is carried along by the mean wind, eddies gradually diffuse the smoke particles over a bigger volume until the density of particles is so small that the puff can no longer by seen. Turbulent diffusion also spreads out the moisture and heat of the atmosphere.

The intensity of turbulence in the atmosphere depends on several factors, but the most important is the stability of the atmosphere—how well the atmosphere arrests vertical displacements of air parcels (see Section 4.5), the roughness of the ground, and the speed of the wind. Figure 4.11 illustrates types of behavior of a smoke plume under laminar and turbulent conditions.

The effect of turbulence on the wind is to cause a transfer of momentum through a much deeper layer of air than would be affected if only molecular diffusion processes were operating. Depending on the speed of the air flow, the roughness of the underlying surface, and the stability of the atmosphere, the drag of the surface on the flow can extend from anywhere between a 300-meter and a 2000-meter elevation.

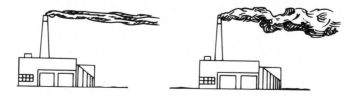

(a)   (b)

Figure 4.11   Effect of turbulence in diffusing smoke: (a) laminar (nonturbulent) flow, (b) turbulent flow.

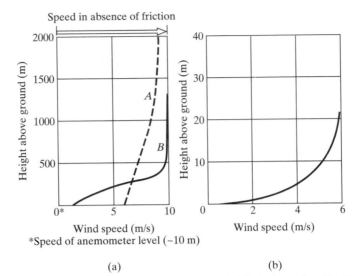

Figure 4.12 Examples of the variation of wind speed with height in the surface "friction layer": (a) A = strong vertical mixing, B = weak vertical mixing, (b) variation of wind speed in the first 40 meters.

On a day when the atmosphere is well-mixed [curve A in Figure 4.12(a)], the surface can be "felt" as high as 2000 meters. In contrast, when the atmosphere is stratified and vertical mixing is suppressed [curve B in Figure 4.12(a)], the surface drag extends upward to only 500 meters or less. The speed of the wind at the anemometer level (usually 5–10 meters above the surface) is only a fraction of the speed in the free atmosphere [Fig. 4.12(b)].

The effect of friction on the horizontal wind that is observed near the ground is illustrated in Figure 4.13. Near the surface, the frictional drag (always opposed to the direction of air motion) slows the wind; the Coriolis force is correspondingly less, and the "steady," or mean, wind is that resulting from a balance of three forces— pressure gradient, Coriolis, and friction. Since the Coriolis force and friction are always at 90° and 180° from the wind direction, respectively, a balance of the three forces can be achieved only if the wind blows at an angle across the isobars. The angle can be 45° or more, but it is usually only 20 or 30°. At higher elevations, the wind speed increases and the direction cuts across the isobars at a smaller angle; the three forces are again in balance as far as the steady wind is concerned. Above the **friction layer** (i.e., where friction becomes negligible) the wind is stronger still and its direction is essentially parallel to the isobars. As mentioned earlier, such a wind, resulting when the only forces acting are the Coriolis and pressure gradient, is called *geostrophic*. The level at which the wind is a close approximation of the geostrophic is normally above 300 meters.

The above discussion of the variation of wind with height and Figure 4.13 assumes that the horizontal pressure gradient does not vary with height. If the horizontal pressure gradient changes with increased height, the wind will change due to

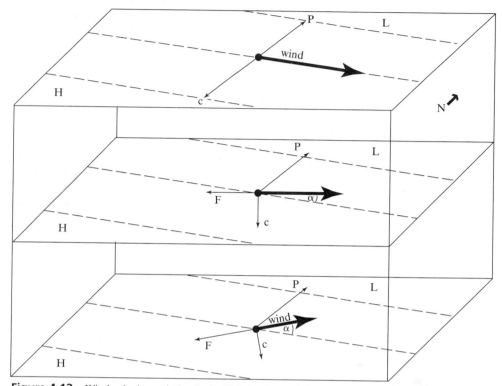

Figure 4.13   Wind velocity variation in the "friction layer" (Northern Hemisphere). Parallel dashed lines are isobars with high pressure to the south. The pressure gradient force is denoted by P; it is balanced by the sum of the Coriolis force (c) and friction force (F). The angle between the wind direction and the isobars is designated by α.

this effect as well as to the decreased frictional drag of the ground. In fact, the horizontal pressure gradient normally does vary greatly with altitude. In middle latitudes the pressure gradient force normally reaches a maximum around 8 km, and it is at this level that the fastest winds (jet stream) occur.

The net result of the horizontal pressure gradient force, the Coriolis force and friction determine the circulation around circular high and low pressure centers near the Earth's surface. The types of flow patterns associated with low pressure areas (**cyclones**) and high pressure areas (**anticyclones**) near the Earth's surface in the Northern and Southern Hemispheres are shown in Figure 4.14. In the Northern Hemisphere, the air is seen to spiral in a counterclockwise sense into a cyclone; it spirals in a clockwise sense out of an anticyclone. In the Southern Hemisphere, the spiral is inward in a clockwise sense for a cyclone, it is outward in a counterclockwise sense for an anticyclone. The approximate locations of high and low pressures can be determined from the observed wind direction at a point by use of *Buys Ballot's rule*: If you stand with your back to the wind, low pressure is to your left, high pressure is to your right (Northern Hemisphere).

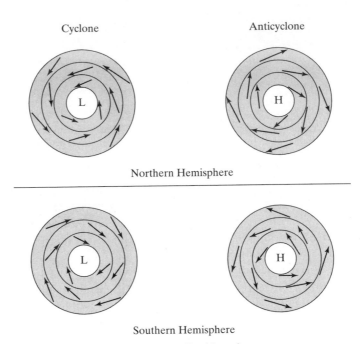

Figure 4.14   Cyclonic and anticyclonic flow near Earth's surface.

4.4   VERTICAL MOTION AND ITS RELATION TO CLOUDS

We have seen how differences in temperature over the globe create thermal circulations, which involve both vertical and horizontal motions. Because of the important role played by vertical motion in producing weather, some further discussion of how vertical motion is produced and how it affects the formation and dissipation of clouds will be given here.

Vertical displacements of air result from "dynamic" causes and from vertical accelerations associated with buoyancy. One dynamic cause is mountains. They act as barriers to horizontal air flow, forcing it to ascend along the windward sides and descend on the lee sides. On a large scale, another important cause of vertical motion is the divergence and convergence of air currents circulating around the great anticyclonic and cyclonic whirls of the atmosphere. Air currents spiraling inward at low levels of cyclones converge; i.e., they move toward the center. Since the horizontal area occupied by a volume of air must therefore decrease with time, the vertical depth must increase. This is illustrated by Figure 4.15(a). Imagine a column of air, having the boundaries shown in the figure, with streams of air spiraling inward toward the center. The *inward* component of the flow would result in a shrinking of the cross-sectional area with time (**convergence**). Since the amount of mass contained in the imaginary cylinder must remain constant, it follows that the depth of the cylinder must increase. Air must therefore move upward within the column. Conversely, in the lowest few kilometers of anticyclones, outward flow everywhere would result in an expansion of the column's horizontal cross section (**divergence**); vertical shrinking would be required to keep the

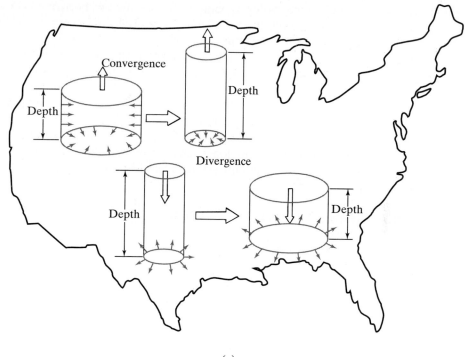

(a)

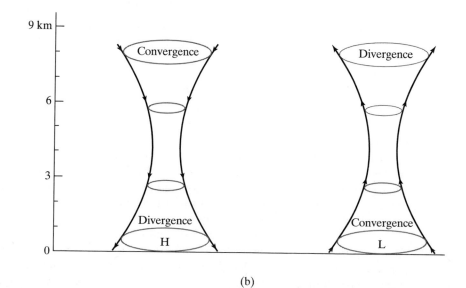

(b)

Figure 4.15 (a) Convergence and divergence of a disk or air. (b) Large-scale patterns of divergence and convergence associated with surface anticyclones (highs) and cyclones (lows).

total volume constant. At levels about 6 to 7 kilometers, horizontal convergence occurs over surface anticyclones, while divergence occurs over surface cyclones, so that a pattern of motion such as that illustrated in Figure 4.15(b) emerges. Note that the vertical scale in this drawing is greatly exaggerated. The diameter of the typical anticyclone or cyclone is greater than 1000 kilometers; thus, the downward and upward flows are not nearly as steep as they appear in Figure 4.15(b). The vertical velocities produced by these large-scale patterns of divergence and convergence are usually not more than a few centimeters per second (1 kilometer per day). However, they are sufficient to set the weather "stage" over large areas. In the absence of other influences, the weather over areas dominated by cyclones tends to be that of widespread cloudiness and precipitation, while that over anticyclonic areas is frequently clear.

4.5 VERTICAL STABILITY AND ITS RELATION TO VERTICAL MOTION AND CLOUDS

In discussing the characteristics of the troposphere (Chapter 1), it was mentioned that convection keeps this lowest layer fairly well stirred, in contrast to the stratosphere, in which there is not very much mixing. Yet, on the average, the temperature in the troposphere is not uniform in the vertical, but rather it decreases at the average rate of 6.5°C/km. Evidently, a well-mixed layer of air does not imply one of constant temperature, at least not in the vertical. The reason for this is linked to the pressure changes that air parcels experience during displacements.

Adiabatic Processes

Air displaced vertically experiences rapid pressure changes; in response to these changes, the volume must increase or decrease. For example, if a parcel of air is forced to descend from an elevation of one kilometer, the pressure exerted on it will have increased by about 20 percent by the time it reaches the surface. In response to such a pressure change, the volume and/or the temperature must also change. An indication of what actually occurs can be obtained from our experience in letting air escape from a tire—as the air expands and the pressure drops rapidly, the air cools. The air cools because some of its heat energy is expended in doing work of expansion. Conversely, if we pump up the tire rapidly, the air warms up because the work we have done in compressing the gas is converted to heat.

The same process takes place when the pressure of an air parcel is rapidly changed by its descent or ascent into the atmosphere. Assuming that the heat loss or gain by an air parcel through conduction, radiation, and mixing with the surroundings is at a slow enough rate during a vertical displacement, the temperature changes can be ascribed mostly to volume changes.

The ideal or theoretical process during which there is absolutely no heat exchange between a mass and its environment is said to be *adiabatic*. Since air usually contains water, and phase changes of water involve latent heat, we distinguish between two different adiabatic processes: (1) A **dry adiabatic** process is one during which there are no phase changes of water (no condensation, evaporation, freezing, or sublimation); (2) A *moist* or *wet adiabatic* process is one during which phase changes *do* occur, and account must be taken of the latent heat.

Since the horizontal pressure gradient is small, and the wind crosses the isobars at a small angle, compression or expansion of air parcels moving in the horizontal is very small. For this reason, we are concerned mainly with volume changes of air parcels when they move up or down.

A dry adiabatic displacement of a parcel of air in the vertical results in a temperature change of 9.8°C per kilometer, or nearly 1°C per 100 meters of elevation (5.5° per 1000 feet). This rate of temperature decrease with height is illustrated in Fig. 4.16 by the solid line. Following the line upward from a temperature of 20°C at the surface (pressure approximately 1013 mb) to an elevation of 7000 m (pressure approximately 410 mb), we see that the temperature of a parcel of air would decrease almost 70°C as the volume of the parcel of air expanded. If the same parcel of air were to return to the surface, its temperature and volume would assume their initial values.

During a wet adiabatic process, the changes of phase of the water contained within a parcel of air experiencing a rapid pressure change cause conversion of latent heat to sensible heat, or vice versa, depending upon whether the parcel is rising or shrinking. When condensation is occurring, the latent heat released raises the temperature of the air parcel; when evaporation is occurring, the latent heat required cools the air. In the example shown in Figure 4.16, if the air parcel at sea level were saturated with water vapor, each kilogram of dry air would contain about 10.7 grams of water vapor; any lifting of the parcel would cause expansion and cooling, and the excess moisture would have to condense. Thus, when the parcel reached an altitude of 2000 meters (pressure ~ 800 millibars), the air would still be saturated with water vapor, but the amount of moisture in vapor form would be less than two-thirds the original value. More than 4 grams of water in each kilogram of air would have condensed, releasing about $4 \times 600 = 2400$ calories of latent heat. As a result, the temperature of the air parcel would be considerably warmer (9°C) than it would have been had the process been dry adiabatic. Further ascent of the parcel would cause

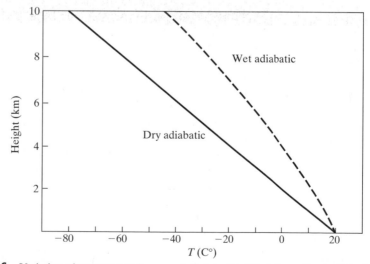

Figure 4.16   Variation of temperature in a parcel of air as it is lifted dry and wet adiabatically from a height of 1 to 10 km. The dry adiabatic lapse rate is a constant 9.8°C/km; the wet adiabatic rate varies from about 6°C/km near the surface to 9.8°C/km at 10km.

more condensation, but because the rate at which condensation proceeds is less when the temperature is low than when it is high, the rate of temperature change during a wet adiabatic process is not constant. In Figure 4.16 this can be seen by comparing the temperature change between 2000 meters and 4000 meters (about 6°C/km) with that between 8000 meters and 10,000 meters (about 9°C/km).

If the "wet" air parcel illustrated in Figure 4.16 were to descend, it would warm at the same rate that it cooled during the ascent, *if* all of the liquid and solid water remained in the parcel. In practice, some of the water drops and ice crystals leave the parcel of air, so that if the parcel later descends, there will be less evaporation and hence a more rapid warming than occurs in the true wet adiabatic process. When some of the condensation products drop out, the process is said to be pseudoadiabatic because it is "irreversible."

Condensation in the atmosphere is produced principally through the cooling of air as it ascends, comes under lower pressure, and expands. Similarly, the dissipation of clouds is usually a sign of descending air. The overall pattern of vertical motion producing a cumulus cloud, such as that shown in Figure 4.17, is upward motion below and within the cloud and downward motion at the edges outside the cloud. (In a well-developed cumulonimbus cloud, the pattern is more complex, as we shall see later.) Even air containing little water vapor (low relative humidity) does not require a great deal of vertical lift to create saturation and then condensation. For example, air starting near sea level with a temperature of 30°C and a dew point of 14°C (relative humidity = 36 percent) will become saturated at about 2000 meters elevation. (The dew point decreases at the rate of almost 2°C per 1000 meters of ascent during a dry adiabatic process. The height at which saturation will be reached is, therefore,

$$\frac{(30° - 14°)}{1°/100\text{m} - 0.2°/100\text{m}} = 2000\text{m}$$

After saturation, both the dew point and temperature decrease at the wet adiabatic rate.)

Vertical Stability

The ability of the atmosphere to produce and sustain vertical currents depends on the atmosphere's "stability." A **stable** atmosphere is one in which **buoyancy** forces oppose the vertical displacement of air parcels from their original levels. An **unstable** condition exists when buoyancy forces promote or abet the vertical displacement of air parcels. A **neutral** state exists when vertical displacement is neither opposed nor abetted by buoyancy forces.

The buoyancy of a parcel of air will depend on its density relative to the environment density at the same level. If a parcel is "heavier" than the medium surrounding it at the same level, it will sink; if it is lighter, it will rise. If its density is the same as that of its surroundings, there will be no "Archimedean force" tending to make it either rise or fall.[†]

[†]Archimedes (born 287 B.C.) used the principle of buoyancy to determine the percent of gold in the crown of Hiero, King of Syracuse.

Figure 4.17 Stages of development of a cumulonimbus over Arizona. (Courtesy of L. Batten, University of Arizona.)

Since we do not normally measure density directly in the atmosphere, it is more convenient to discuss stability in terms of temperature, rather than density. From the equation of state (Eq. 1.4), we know that any fixed pressure the density is inversely proportional to the temperature. Therefore, at any given pressure level, we can substitute temperature for density in the statements on buoyancy in the previous paragraph: a parcel of air that is warmer than its surroundings will rise; one that is colder than its surroundings will sink; and one at the same temperature as its environment will not experience a force in either direction.

We have seen from the discussion of adiabatic processes that the temperature of an air parcel displaced vertically changes at a *fixed* rate—1°C/100 m if there is no water-phase transition and at some lesser rate if there is condensation or evaporation. Evidently, then, whether a vertically displaced parcel of air is warmer or colder than its environment at any particular point along its path will depend on the vertical distribution of ambient atmospheric temperature. The rate at which the temperature decreases vertically in the atmosphere is called the **lapse rate**. The *average* lapse rate in the troposphere is 6.5°C/km, but the *actual* lapse rate at any given point may vary greatly from this average value. It is the actual lapse rate that determines the stability of the atmosphere. We will illustrate three important conditions of atmospheric stability by the three hypothetical lapse rates shown by soundings (vertical temperature profiles) *A*, *B*, and *C* in Figure 4.18. The stability of the atmosphere depends on the relationship of the actual lapse rate to the dry adiabatic and wet adiabatic lapse rates (shown in Figure 4.16), so these are also plotted in Figure 4.18.

An atmospheric layer in which the *actual* lapse rate is *less* than the *wet adiabatic* lapse rate is *absolutely stable* and is illustrated by temperature sounding *C* in Figure 4.18. The atmosphere is said to be absolutely stable because no matter whether a parcel of air is saturated or unsaturated, an upward vertical displacement will cause the parcel of air to cool at a faster rate than the temperature of its environment cools. For example, if an unsaturated parcel of air at the surface and with a temperature of 20° C were displaced upward, its temperature would fall at the rate given by the dry adiabatic curve in Figure 4.18. It would quickly become cooler than the environment, which cools at the rate given by curve C in Figure 4.18. As it became cooler it would become heavier than the surrounding air and buoyancy forces would cause the parcel to sink back toward its original level. A saturated parcel displaced upward would cool at a slower rate than the unsaturated parcel (following the dashed wet adiabatic curve in Figure 4.18), but it too would become cooler and heavier than the environment and also be forced back downward. Thus, the temperature sounding *C* in Figure 4.18 represents an absolutely stable atmosphere.

An **absolutely unstable** atmosphere is one in which the actual lapse rate is *greater* than the dry adiabatic lapse rate; this condition is represented by sounding *A* in Figure 4.18. In an absolutely unstable atmosphere, both unsaturated and saturated parcels of air that are displaced upward become progressively warmer and lighter than their environment and are accelerated upward, away from their initial position.

An interesting, and fairly common, situation occurs when the atmospheric lapse rate is greater than the wet adiabatic lapse rate but less than the dry adiabatic lapse rate. This condition, which is call **conditionally unstable**, is illustrated by sounding *B* in Figure 4.18. Under conditionally unstable conditions, an unsaturated parcel of air

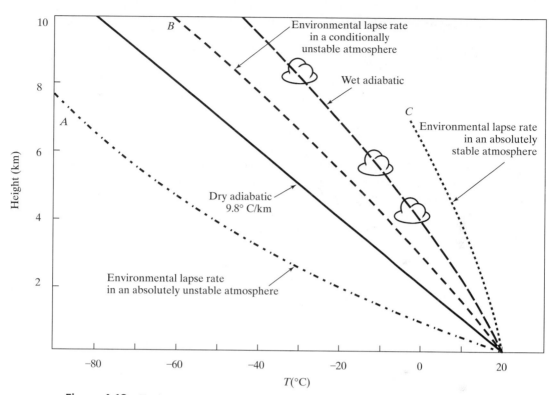

Figure 4.18   Environmental lapse rates in an absolutely unstable atmosphere (sounding *A*), a conditionally unstable atmosphere (sounding *B*), and an absolutely stable atmosphere (sounding *C*). The dry adiabatic and wet adiabatic lapse rates are shown for comparison.

which rises and cools at the dry adiabatic rate will become cooler and heavier than its environment and will be forced back downward toward its initial position, a stable situation. However, a saturated parcel of air which rises and cools at the wet adiabatic lapse rate will become warmer and lighter than its environment, and be accelerated away from its initial level, and unstable situation.

Since the atmospheric lapse rate in the real atmosphere is often between the dry and the wet adiabatic lapse rates, we can see the importance of the amount of water vapor in the atmosphere in promoting or resisting vertical displacements. If displaced upward for whatever reason, moist parcels of air will cool and reach saturation much more quickly than dry parcels of air. After saturation is reached, these parcels will cool at the wet adiabatic lapse rate and become warmer than their environment, continue to rise, producing more condensation, clouds and eventually precipitation. In contrast, dry parcels of air will become cooler than their environment and will be inhibited from further rising. Thus condensation of water with its release of latent heat is an important factor in inducing vertical currents of air. We see here the basis for an earlier statement that a significant portion of the energy that drives the atmosphere occurs through the evaporation-condensation cycle of water.

Changes in Stability

Lapse rates vary considerably in space and time. Increase of the lapse rate in a layer of atmosphere results from warming of the lower part of the layer and/or cooling of the upper part. Conversely, a decrease of the lapse rate is produced by cooling of the lower portion and/or warming aloft. Some of the possible causes for different rates of temperature change in a layer follow.

Differential Heat Transport.   If the air aloft is being replaced by warmer air brought in by the winds (**warm advection**), while in the lower portions colder air is being brought in (**cold advection**), the stability of the layer will increase. The opposite, cooler air coming in at high levels with warmer air near the bottom of the layer will increase the instability of the layer.

Surface Heating or Cooling.   This can occur in either of two ways: (1) It occurs when air moves over a surface that is either colder or warmer than itself. For example, air moving from an ocean over a warm continent may cause enough instability to set off showers. In the winter, warm air from the Gulf of Mexico streaming northward over the central and eastern United States sometimes is cooled sufficiently by the cold surface to produce widespread areas of fog and low layers of stratus clouds. (2) When air over a surface that is losing or gaining heat through radiation, the air in contact with the surface will cool or warm. The cooling of the air near the surface on a clear calm night frequently leads to the creation of a temperature **inversion** (i.e., a layer of *increasing* temperature with height).

Radiative Cooling Aloft.   Clouds are quite effective blankets, retaining the heat of the air below them. The loss of heat at the *tops* of clouds, though, can lead to instability within the cloud layer.

Vertical Displacements of Layers.   When an entire layer of air sinks (*subsidence*), the difference in the percent compression between the bottom and top of the layer leads to a greater warming of the upper portion than of the lower portion of the layer, and therefore greater stability. Inversions created in this manner are known as *subsidence inversions*. Conversely, when a layer of air is lifted *dry* adiabatically, its stability decreases. However, if part of the layer becomes saturated during the ascent, the situation changes. If the upper portion becomes saturated during the ascent, the situation changes. If the upper portion becomes saturated before the lower, the stability will be increased, because the upper zone will cool at the lesser wet adiabatic rate during the ascent. But if the lower portion becomes saturated first, it will cool slower than the upper part and the instability within the layer will be enhanced. Thus, instability is more likely to occur in a layer of air if the bottom of the layer is more nearly saturated at the onset of lifting.

Stability and Clouds

Cumuliform clouds are associated with instability and strong vertical motion. The convective ascent of air in such clouds appears to occur in bursts or bubbles, somewhat like those that form in boiling water. Each successive bubble, having dimensions of up to a few kilometers in the horizontal and a few hundred meters in the vertical, rises, expands, and cools; but as the bubble ascends through the atmosphere it is "eroded" or mixed with the drier surrounding air and gradually loses its identity (Figure 4.19). In

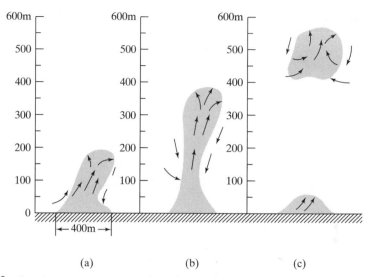

Figure 4.19 Development of convective bubbles producing a "thermal."

this way, puffs of cumulus may form, gradually disappear, and perhaps be replaced by new puffs. Horizontal winds may carry each bubble downstream from the surface point where it was created. Strong, turbulent flow tends to cause rapid erosion.

However, if the surrounding atmosphere is moist and the horizontal winds and turbulent mixing are not too great, previous bubbles may still remain while new bubbles are being formed. Then, each subsequent bubble will be able to ascend to ever greater height, causing the cumulus cloud to develop vertically. With increased instability, enhanced by the release of latent heat within the cloud, the convection becomes more continuous and a well-defined thermal circulation, such as occurs in the towering cumulonimbus of a thunderstorm cell, may develop. The rapidity with which such a convective cell can develop is illustrated in the photographs of Figure 4.17.

Columns of rising air are sometimes called **thermals**. Glider pilots learn to seek out thermals and then try to remain within their boundaries so they can be carried upward. All around the thermal there is a compensating downward flow of air.

KEY TERMS

acceleration	dry adiabatic process	molecular viscosity
anticyclone	eddies	neutral
balance of forces	friction	Newton's first law
buoyancy	friction layer	Newton's second law
centrifugal force	geostrophic equilibrium	pressure-gradient force
centripetal force	geostrophic wind	stable
cold advection	gravitational force	thermals
convergence	hydrostatic balance	turbulence
Coriolis force	hydrostatic equation	unstable
cyclone	inversion	warm advection
divergence	isobars	wet adiabatic process
eddy viscosity	lapse rate	

PROBLEMS

1. In terms of density, explain how a balloonist is able to ascend or descend at will. Before light gases such as helium were available, balloonists inflated their balloons with hot air. What does this show about the effect of temperature on density and thus buoyancy? Why does smoke rise? Explain how portions of the atmosphere acquire buoyancy.

2. How would you classify the mean or standard lapse rate in the atmosphere as far as stability is concerned?

3. Compute the height of the base of a cumulus cloud formed by a thermal if the surface temperature and dew point are 85°F and 49°F, respectively.

4. If the Earth were not rotating, what would be the direction of the horizontal wind with respect to the isobars? What force would keep the air from increasing its speed ad infinitum?

5. The plane of a freely suspended (Foucault) pendulum rotates about the local vertical at all latitudes except the equator. At what rate does it rotate? (Review the discussion of the Coriolis force.)

6. Can there be a geostrophic wind at the equator? Explain your answer.

7. Explain the reasons for the horizontal and vertical pressure gradients in terms of density variations within the atmosphere.

8. Plot a graph having as the abscissa, temperature over the range of +30°C to -55°C, and as the ordinate, height over the range 0 to 16 kilometers. On the right-hand vertical scale, indicate the standard pressure in the vertical, as determined from Appendix 2 and Figure 1.5. Plot the following three temperature-pressure soundings measured in three different air masses by connecting consecutive points in each sounding with straight-line segments. Label each curve. Draw several straight sloping lines on the chart to illustrate the rate at which temperature changes with height during a dry adiabatic process, one line starting at 30°C and sea level, another at 0°C and sea level, and a third at -30°C and sea level.
   a. Identify layers in the three soundings that exemplify absolutely stable, absolutely unstable, and neutral stratifications.
   b. Identify all layers containing either an inversion or an isothermal lapse rate.
   c. Where would you say the tropopause is located in the three soundings?

Air Mass Pressure (mb)	Tropical Temp. (°C)	Polar (summer) Temp. (°C)	Polar (winter) Temp. (°C)
1000	27	13	-31
950	26	13	-32
900	25	9	-32
850	22	4	-30
800	20	-1	-30
700	13	2	-28
600	6	-8	-31
500	-1	-17	-38
400	-13	-32	
300	-28	-50	
200	-50	-50	
150	-51	-50	

5

Atmospheric Circulations and Storms

5.1 SCALES OF MOTION

In the last chapter, it was shown that pressure differences produce the basic force that drives the winds. However, the picture of the air flow is made complex by the Earth's rotation, frictional drag and turbulence, mountain obstacles, and, perhaps most of all, the extremely variable character of the Earth's surface and the incessant changes of the state of water in the air. To simplify the analysis of the enormously complex patterns of "eddies within eddies" that exist in the atmosphere, it is convenient to categorize circulation systems according to size. Almost every size is represented in the atmosphere—everything from the very small whirls that kick up the dust on a road to enormous circulations that have horizontal dimensions of thousands of kilometers. All of these different sizes—or **scales of motion,** as they are called—are interdependent. For example, an eddy produced by a hill might not occur unless there were a prevailing wind due to a circulation of much larger size.

An instantaneous snapshot of the winds of the entire atmosphere would present an extremely chaotic view of the flow. The complex distribution of forces producing such flow would make prediction an impossible task. To achieve some order, a type of filtering, according to size of flow elements, must be applied. This is accomplished by a system of averaging.

The very small-scale eddies or whirls that cause a flag to oscillate rapidly or branches of a bush to sway for periods of perhaps only a few seconds can be eliminated by averaging the observed wind velocity over periods of several minutes. If one were to average the wind velocity over an entire day, then wind oscillations having periods of much less than a day would disappear from the record. Meteorological observations are averaged over time and space to isolate the various sizes of atmospheric motions. The analyst of the weather maps that are published in the newspapers or shown on television applies an averaging process—a smoothing of isobars—that eliminates most irregularities smaller than about 100 kilometers.

Most routine meteorological measurements are made in such a way that very small eddies are eliminated. Most anemometers and thermometers do not react to small, high-frequency changes. Observations are so widely spaced in time and area that most must be considered averages over horizontal distances of tens of kilometers and vertical distances of tens of meters. Even such relatively large phenomena as thunderstorms and tornadoes often fall through the "mesh" of the usual weather station network.

The scales of atmospheric motions can be classified as shown in Figure 5.1. On the thermometer **microscale** are small, short-lived eddies, often referred to as turbulence, that are strongly affected by local conditions of both terrain roughness and temperature. These eddies are very significant as dispersers of pollutants. The Coriolis force is not significant on the microscale. At the large end of the microscale are tornadoes and waterspouts.

The **mesoscale** includes a variety of phenomena of intermediate horizontal size such as land-sea breezes, thunderstorms, and squall lines. Coriolis forces may play an important role in the larger phenomena in this class, such as sea breezes.

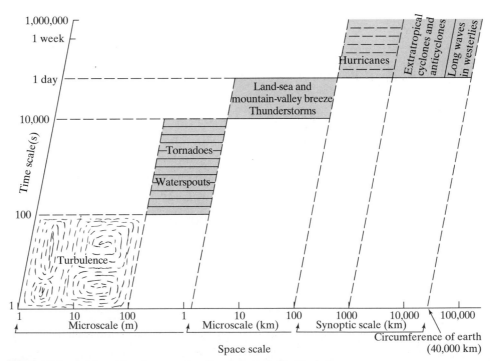

Figure 5.1   Horizontal and temporal scales of atmospheric circulations.

The cyclones and anticyclones that are largely responsible for the day-to-day weather changes belong to the large or **synoptic* scale**. These systems persist for days or weeks and Coriolis forces are very significant. At the largest end of the synoptic scale, sometimes called the **planetary scale**, are features of the atmospheric circulation that persist for weeks or months. Long waves that exist in this flow move very slowly, or not at all, across the Earth. These play an important role in determining the seasonal characteristics of the weather. Coriolis forces are very important on this scale.

The wind observed at any place can then be thought of as a composite of several different scales of motion. For example, the synoptic flow patterns are associated with the large features of the Earth's surface and distribution of heat: continents and oceans, extensive mountain ranges, latitudinal variations of insolation. The various scales of motion can also be characterized by the magnitude of the vertical motion associated with each. Synoptic-scale motion is mostly in the horizontal. The vertical displacements attributable to the very large circulation features are no more than a few cm/s; even in the great cyclonic storms that regularly affect the middle and high latitudes, average vertical displacements are only of the order of 50 cm/s (1 mi/h). In the smaller, more intense mesoscale circulations, the vertical velocities are more comparable to the horizontal velocities; for example, in a thunderstorm, the vertical motion is often 10 m/s (22 mi/h) and can reach 30 m/s or more. Winds of the microscale are generally much weaker than those of the larger sized motions, but the vertical motions are very nearly equal to those in the horizontal. However, microscale motions, unlike those of the mesoscale, occur principally in a rather shallow layer adjacent to the Earth's surface.

5.2 THE GENERAL CIRCULATION AND MONSOONS

At the beginning of Chapter 3 attention was drawn to certain similarities between the heat engine and the atmosphere. This analogy can be demonstrated in yet another way. Just as the motor of an automobile functions through the turning of wheels and gears of various sizes, the total kinetic energy of the atmosphere is partitioned among circulations of varying dimensions. Furthermore, in the mechanical engine, the larger the mass of the rotating wheel or gear, the greater the power required to turn it. In the atmosphere, also, the amount of energy required to initiate and maintain a circulation pattern is, for the most part, directly proportional to the mass of air associated with the circulation.

The wind system which contains the greatest mass of air is called the **general circulation**. Driven by the energy received from the sun, the system transports the air from the equatorial regions toward the poles, and maintains a return flow of cold air from polar to tropical latitudes. It determines, in large measure, the broad pattern of climates of the Earth. Within this large-scale global flow are embedded the smaller circulations. These smaller circulations—"perturbations" of the worldwide flow—are

---

*In meteorology, *synoptic* means "coincident in time." Thus a synoptic weather map shows conditions as they were at a particular time. The word "synoptic" has come to denote large-scale as opposed to small-scale atmospheric patterns.

responsible for the transient, short-period variations of atmospheric conditions, i.e., the weather. Here, we shall examine the structure and causes of various sizes of atmospheric circulation patterns, starting with the largest scale.

The General Circulation

The general circulation of the atmosphere is a time-averaged flow of air over the entire globe. It is determined by averaging wind observations over long periods of time—usually twenty years or more. To isolate the seasonal variation of the general circulation induced by the Earth's revolution about the sun, the averaging is sometimes done separately for each season of the year. In any case, this long-period averaging tends to eliminate the smaller circulations, as explained at the beginning of this chapter.

If the Earth were not rotating and if the surface were homogeneous, solar heating at the equator would cause the air in that region to rise and flow toward the poles. As it was transported poleward, not only would the air become cooler and tend to sink toward the surface, but the convergence of the meridians of longitude would force the air to "pile up" before it reached the pole. These effects would induce a return circulation near the Earth's surface from polar regions to the equator. In practice, however, this simple thermal circulation pattern between the poles and the equator is greatly modified by the rotation of the Earth and by the nonuniform properties which are characteristic of its surface. Instead of a single circulation cell from equator to pole in each hemisphere, there are three latitudinal circulations, and there are also important longitudinal variations around each hemisphere.

The picture of the general circulation can be simplified somewhat by averaging the observed winds along each latitude, thus eliminating the longitudinal variations. Figure 5.2 is a schematic representation of the results. The horizontal flow at the Earth's surface is shown in the center of the diagram; the net meridional circulation, at the surface and aloft, is depicted around the periphery. The component of the flow along meridians (north/south) has a speed, on the average, of less than a tenth of that along latitude circles (west/east), indicating the importance of the effect of the Coriolis force on this largest of scales.

Within the equatorial region are the **doldrums**, or intertropical convergence zone, a belt of weak horizontal pressure gradient and consequent light and variable winds. Here, also, is the region of maximum solar heating, where the surface air rises (as shown by the vertical circulation at the edge of the diagram) and flows both northward and southward toward the poles. This poleward flow at high levels is acted upon by the Coriolis force, turning the wind to the right in the Northern Hemisphere and to the left in the Southern Hemisphere. Thus, in both hemispheres, the poleward-flowing air becomes a west wind and at an average latitude of about 30°, reaches a maximum speed which may exceed 160 km/h (100 mi/h). These are the **jet streams**, which are discussed in more detail later on.

At about 30° north and south latitudes, some of the air descends again toward the surface. This is the region of the **horse latitudes** (so-called because Spanish sailing vessels carrying horses to the Americas were occasionally becalmed in these areas of light winds and many of the animals had to be thrown into the sea because of the lack of food). Since the air is generally descending in these zones, there is little cloudiness or precipitation, and it is here that most of the world's great deserts are found.

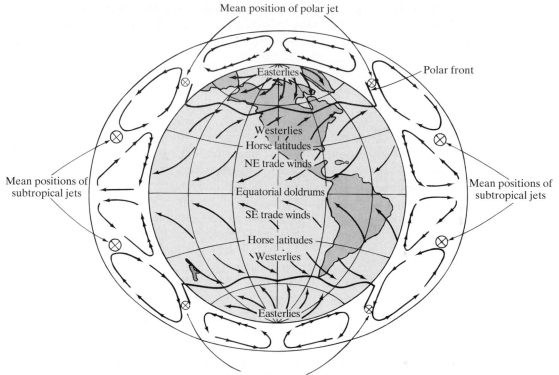

Figure 5.2 Schematic representation of the general circulation of the atmosphere. Double-headed arrows in cross section indicate wind component from the east.

Between the doldrums and the horse latitudes are wide belts where a portion of the original equatorial air, having been cooled by its journey northward and dried out by its descent to the surface, returns again to the tropics. However, it does not flow directly southward, but is deflected by the Coriolis force, so that the wind moves from the northeast in the Northern Hemisphere and from the southeast in the Southern Hemisphere. These are the remarkably persistent **trade winds**, which obtained their name because of the important role they played in opening up the New World when ships were dependent upon sails.

From the horse latitudes, some of the descending air moves poleward at low levels, but the flow is deflected by the Coriolis force and the winds have a westerly component. These are the **prevailing westerlies**. During the days of sailing ships, they provided the motive power for vessels returning from North America to Europe. As the warm poleward-moving air reaches a latitude of from 40° to 60°, it encounters a cold flow from the pole. As a result of this encounter, a boundary, known as the **polar front***, is formed between the two masses of air. Here, the warm, light air from the horse latitudes is forced to rise over the cold, dense air from the pole, and a portion of the warm air returns at high levels toward the equator.

*We shall discuss the polar front in more detail later in this chapter.

Poleward of the polar front are the **polar easterlies**. These winds bring the cold arctic and antarctic air from the polar regions toward the polar front, where they are warmed by heating from the warmer surface below (and also, in individual storms, by mixing with the warmer air on the other side of the front). The air thus rises and returns toward the poles as a westerly flow aloft.

From this description of the general circulation of the atmosphere, we see that there are two primary zones of rising air—in the tropics and in the region of the polar front. As might be expected, it is here that the principal areas of precipitation are found. Complementing these regions are the zones of descending air—in the horse latitudes and near the poles. Here, the precipitation is relatively light. We have already noted that the main deserts of the world are found in the horse latitudes; precipitation in the polar regions is also very small. However, because of the very low rate of evaporation in polar regions, ice remains on the ground for long periods of time.

The preceding discussion represents an average or *mean* description of the atmospheric circulation for the entire year as a function of latitude only. The patterns of mean sea-level pressure during January and July in the Northern Hemisphere are shown in Figure 5.3 and give some idea of how the circulation varies over the surface of the Earth during the year. Note that the subtropical high-pressure belt, associated with the accumulation of air in the horse latitudes, is not continuous around the hemisphere either in winter or summer, but rather is broken up into cells over the Atlantic and Pacific Oceans. These two cells are especially well defined in summer and are displaced several degrees of latitude farther northward in summer than in winter.

Since the winds blow clockwise around high pressure in the Northern Hemisphere, the *eastern* periphery of each cell is under the influence of relatively cool, dry northerly flow. Thus, the coastal areas of southwestern North America and Europe are favored by generally pleasant rainless summers. In the interior of these regions (including north-

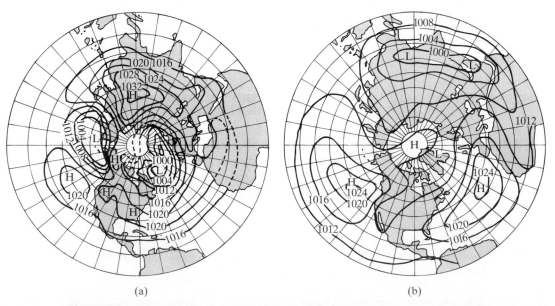

(a)   (b)

Figure 5.3   Normal sea-level pressure in the Northern Hemisphere; (a) January, (b) July.

ern Africa) are the major deserts. The *western* periphery of each high-pressure cell is associated with warm, humid flow from the tropics. Accordingly, locations such as the southeastern United States, as well as Hawaii, the Philippines, and southeastern Asia, are typically warm, with high humidity and frequent summer showers.

Near the latitude of the polar front, where the relatively warm, humid prevailing westerlies meet the cold polar easterlies, are two low-pressure centers. These are well defined in winter, but they almost disappear in summer. Because of their locations, they are termed the **Aleutian low** (in the Pacific) and the **Icelandic low** (in the Atlantic). They represent semipermanent "centers of action," where the major mid-latitude storms develop their greatest intensity.

In the interior of the North American and Asian continents, the low temperatures of winter result in increased density of the air at low levels and produce the cold high-pressure systems noted in those areas. However, in summer, temperatures are high, the air is less dense, and warm low-pressure systems prevail.

Figure 5.4 shows the corresponding flow at 500 millibars (about 18,000 feet) during summer and winter in the Northern Hemisphere. At these levels, westerly winds dominate the region poleward of the horse latitudes, but are more intense in winter than in summer, as evidence by the closer spacing of the winter contours. The geographic center of the flow (lowest contour height) is not generally located at the pole, but it is displaced some distance away. In the Northern Hemisphere, the center of the lowest 500 mb contour is located over western Greenland, with a secondary center over eastern Siberia. These centers are the upper-level portions of the Icelandic and Aleutian lows.

Because the oceans in the Southern Hemisphere cover a significantly larger area than the oceans in the Northern Hemisphere, the temperature is much less affected by fluctuations due to continental influences. Thus, the pressure patterns, and therefore the large-scale winds, are much more symmetrical around the pole in the

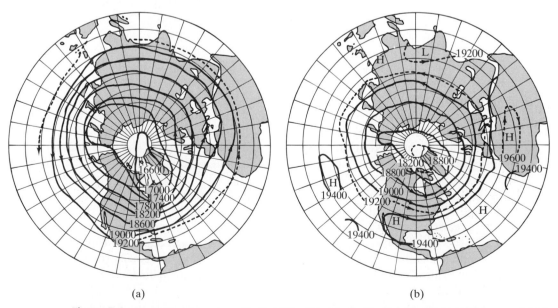

(a)   (b)

Figure 5.4 Normal height contours (feet) at 500 millibars in the Northern Hemisphere; (a) January, (b) July.

Southern Hemisphere than in the Northern Hemisphere. Otherwise, the upper-level flow is similar, being consistently from the west throughout the region from about 30° latitude to near the pole.

Although the existence of jet streams had been postulated by theory much earlier, they were not actually observed until 1946, when high-flying military aircraft encountered unexpected strong head winds against which they could make but little progress. Jet streams are narrow bands of high-velocity winds that meander, like great rivers, around each hemisphere at elevations extending from 4 or 5 kilometers to above the tropopause. The cores of these "rivers" of air are usually about 100 kilometers wide and 2 or 3 kilometers deep, and they flow at speeds as much as 150 km/h faster than the air on either side of them. The location and intensity of jet streams change from day to day throughout the year, but they are associated with zones of strong horizontal temperature change and therefore follow closely the oscillations in position and strength of the polar front. In addition to the circumpolar jet stream, which is normally found between 35° and 60° latitude, a "subtropical jet stream" occurs in the horse latitudes at very high elevations (9—13 kilometers). The subtropical jet stream does not meander over the range of latitudes that the circumpolar jet stream does.

Theory of the General Circulation

Many detailed observational studies indicate that the general circulation is considerably different from that which might be expected for a uniform nonrotating Earth. Clearly, a single thermal cell does not extend from equator to pole. The meridional (south-north) component that would be expected in a simple thermally driven circulation is present only between the equator and 30°. There may also be a much weaker and less persistent polar cell over polar regions. But certainly between these two zones, and perhaps over the entire Earth poleward of 30°, there is no organized thermal circulation; rather, only frequent large eddies (cyclonic disturbances) that intermittently transport heat and momentum between the tropical cell and the polar regions. Note from Figure 5.2 that the mean flow within the middle-latitude cell shows a mean meridional motion that is actually the opposite of what a thermal circulation should have.

A study of the **angular momentum** of the atmosphere indicates that air moving poleward from the tropics would have an excessive westerly speed if there were not some mechanism by which its momentum were reduced. This can be seen by recourse to a basic principle of physics. The angular momentum of any body is given by $m\omega r$, where m is the mass of the body (in this case a parcel of air), ω is the angular velocity of the body, and r is the "spin radius" (i.e., the perpendicular distance of the body from the Earth's axis).

Now if there are no torques (forces that cause turning, such as that applied to a pipe by a wrench), the angular momentum of the air parcel will not change. But if a parcel of air moves from a lower to a higher latitude at a constant distance from the Earth's surface, its distance from the Earth's axis (r) will decrease. Thus, if its angular momentum and its mass (m) remain constant, the angular velocity (ω) must increase. (This is like the skater who makes himself spin faster by pulling his arms in toward his body, thus concentrating his mass near the spin axis.) Because the Earth's angular velocity is the same at all latitudes, if the parcel were originally spinning at the same rate as the Earth's surface, it will be spinning at a faster rate than the underlying surface when it arrives at higher latitudes. This means that as an air parcel moves from the equator toward the pole, it will have acquired an additional speed from the west, relative to the Earth. As an example, a parcel of air displaced from the equator to 60° latitude would acquire a west-to-east speed of about 830 km/h (515 mi/h)! Since such speeds are far greater than those ever observed, some mechanism must be responsible for slowing down the air which is transported from the equator to higher latitudes.

These mechanisms are the cyclonic disturbances, or storms, of middle latitudes. The air in the easterly trade winds, slowed by the friction of the Earth's surface, has its westerly angular momentum increased as it moves toward the equator. This acquired angular momentum, transported poleward by the tropical cell, is gradually absorbed by the great cyclonic storms of middle latitudes and, in turn, is dissipated at the Earth's surface through friction. Thus, these storms embedded in the westerly flow of middle latitudes dissipate the excess momentum, much as the small turbulent eddies near the Earth's surface diffuse a high concentration of smoke in the air.

Monsoons

A large-scale example of a thermal circulation is the **monsoon**. Derived from the Arabic word for season, it refers to a wind circulation that is seasonal in character. During the winter, when continents are colder than the oceans, air flows outward from the continents; during the summer, when the continents are warmer than the oceans, the flow is inward.

During the fall and winter, the continents in middle latitudes cool off faster than the surrounding oceans. The colder air over the continents sinks and spreads out, flowing from the high pressure over the continents to the lower pressure over the oceans. Because of the Coriolis force, the winds flow outward in an anticyclonic fashion. In the spring and summer, the land mass of the continents heats up faster than the oceans and the air over the continents becomes warmer than the air over the oceans. This warmer air rises, and the moist cooler air over the oceans flows inward to the continents, acquiring a cyclonic flow because of the Coriolis force.

Figure 5.5 shows the mean sea-level pressure and the general flow of air near the surface for January and July. The monsoon circulations described previously are most evident over Asia. In January, Asia is dominated by a massive cold high-pressure system (anticyclone) and the low-level air spirals anticyclonically away from the continent. In July, a warm surface low pressure system (cyclone) dominates over Asia and the low-level winds flow cyclonically inward toward the center of the continent. Because the wintertime flow over Asia is cold, dry, and sinking, the Asian winters are very dry. In summer, however, the converging warm, moist low-level air over Asia produces copious amounts of rainfall. Hence, the word *monsoon* has come to be synonymous in many people's mind with torrential rainfalls, although strictly speaking *monsoon* refers to any seasonally varying circulation.

The most intense monsoons are those produced by the large Asian land mass. The climate of southern Asia, protected from the north by the towering Himalayas, is largely determined by the monsoon. In summer, the southerly winds over the northern Indian Ocean deposit the heaviest rainfall in the world along the southern Himalayan

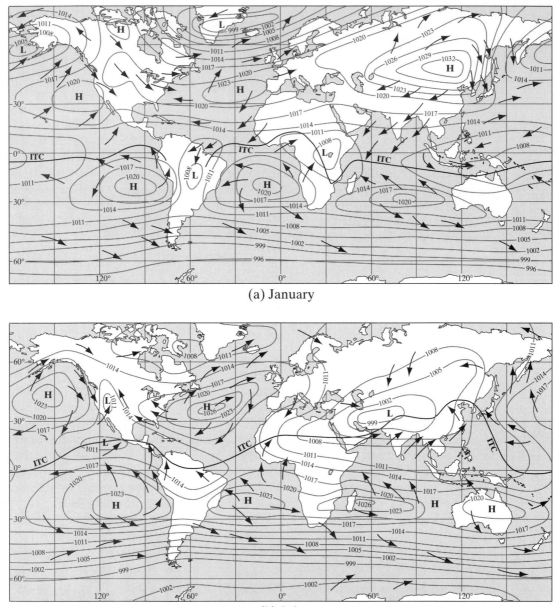

Figure 5.5 Average surface pressure and associated global circulation for (a) January and (b) July. (Source: Lutgens/Tarbuck, *Atmosphere*, 6th Edition, 1995, Prentice Hall.)

slopes. In winter, the prevailing northeast winds are dry and there is little rain. The onset and duration of the summer monsoon are of great significance to agriculture in southern Asia. The geographic distribution, intensity, and duration vary considerably from year to year, and they still cannot be forecast with any degree of certainty.

North America also experiences a monsoon circulation, although it is not near-ly as strong as the Asian one and tends to be obscured on a daily basis by migrato-ry cyclones and anticyclones. On the average, however, as shown in Figure 5.5, the sea-level pressure is lower over North America in the summer than in the winter and the low-level winds tend to blow from the northwest in winter and from the southeast in summer. The strong southeast flow of very warm and moist air in the eastern United States is responsible for the hot, muggy summers in this region. This flow is associated with the large Bermuda High, which reaches a maximum intensity in the summer.

Another feature of the North American monsoon is the northward flow of trop-ical air over the normally arid highlands of the southwestern United States. This un-stable air as it is lifted over the mountains often produces heavy thunderstorms and flash floods in this region.

5.3   CYCLONES AND ANTICYCLONES, AIR MASSES, AND FRONTS

Cyclones and Anticyclones

Embedded in the great circumpolar vortex of the general circulation are the **cyclones** and **anticyclones** of middle and high latitudes. These smaller scale vortices tend to be masked by the averaging process used to analyze the general circulation, but an in-spection of the flow for an individual day reveals their existence. For example, Fig-ure 6.1 shows an intense cyclone centered over Georgia. As we shall see in Chapter 6, record-breaking heavy snowfall produced by this storm paralyzed the East Coast. Southwest of the cyclone, an anticyclone or high was centered over Texas, bringing clear and cold weather to the South. Compared to the size of the circumpolar whirl (about 10,000 kilometers in diameter), cyclonic and anticyclonic eddies, such as those in Figure 6.1, range from 1000 to 4000 kilometers in diameter.

The flow at 500 mb (about 5 km) at 0700 EST, March 13, 1993, is shown in Fig-ure 4.4. The solid lines in Figure 4.4 depict the height of the 500-mb pressure surface above the ground. Height contours on a constant pressure surface are analogous to isobars on a constant height surface, so winds tend to blow parallel to the height con-tours (see Appendix 4 for an explanation of the plotting model for the 500-mb chart).

A comparison of the surface [Figure 6.1] and upper-air [Figure 4.4] charts shows that the flow and pressure patterns aloft are somewhat simpler than those at the surface. The circulation is smoother and more wavelike aloft, with the basic flow being westerly over much of the United States. However, the simple circumpolar vortex that is found in the average flow at 500 mb [see Figure 5.4(a)] is not found on individual days. Rather, the simple pattern is perturbed by traveling waves that are intimately connected with surface cyclones and anticyclones. The troughs of the waves are associated with surface cyclones, while the ridges are associated with an-ticyclones. For example, the surface New England cyclone is associated with the upper-level trough that extends from Pennsylvania north-westward into central Canada and Alaska. We will see in a later section how these upper-level disturbances cause surface cyclogenesis.

A comparison of Figures 6.1 and 4.4 illustrates the concept shown in Figure 4.3. The anticyclone over Texas consists of cold air in the low levels. As illustrated in Figure 4.3, such cold highs weaken with increasing elevation, and therefore no evidence of the high is found at 500 mb. In contrast, the northwest quadrant of the surface cyclone contains cold air; therefore the cyclonic circulation increases with elevation in this region. Finally, we note the cold air aloft is associated with low heights, while warm air aloft is accompanied by high heights [Figure 4.4].

The large cyclonic and anticyclonic eddies that move in a general easterly direction around each hemisphere dominate the flow over much of the Earth between about 30 degrees and about 75 degrees latitude. As was mentioned earlier, they are more significant transporters of heat and momentum between low and high latitudes than is the mean meridional (south-north) flow of the general circulation. This efficient transport occurs because southerly winds around the eastern semicircle of cyclones carry warm air northward, while on the western semicircle northerly winds transport cold air southward. The opposite circulation occurs around anticyclones, so that both pressure systems act as efficient heat exchangers between low and high latitudes.

The cyclonic whirls are the "storms" of middle latitudes. In the temperate latitudes they produce much of the winter precipitation. Around their low-pressure centers, the air circulates in a counterclockwise direction in the Northern Hemisphere and in a clockwise direction in the Southern Hemisphere. The masses of air that circulate around them are generally heterogeneous with respect to temperature and moisture, having come from different geographical areas; as a result, there exist sharp transition zones separating warm, humid air from cold, dry masses. These storm systems go through a complex life cycle which will be discussed in more detail later in this chapter.

Winds in the anticyclone circulations blow clockwise around their high-pressure centers in the Northern Hemisphere and counterclockwise in the Southern Hemisphere. Within these whirls, the air is slowly subsiding at the rate of 10–15 cm/s and "fair weather" generally prevails. The air masses of which they are composed are generally homogeneous with respect to temperature and moisture.

Air Masses

An **air mass** is a huge body of air, extending over thousands of kilometers, within which the temperature and humidity change gradually in the horizontal; i.e., there are no sharp horizontal changes of temperature or humidity. Air masses are created principally within the anticyclonic flow of the subtropical and polar high-pressure belts. The air circulates slowly in these systems over surfaces of fairly uniform properties and gradually acquires thermal and moisture characteristics representative of these surfaces. For example, the air flowing around the semipermanent Atlantic anticyclone very quickly acquires the warmth and moisture of such water bodies as the Caribbean and Gulf of Mexico. Cold air masses, such as those that form over the frozen surfaces of northern Canada and Siberia in winter, take somewhat longer to form, but under fairly stagnant conditions horizontal homogeneity can exist to a 3-kilometer or 4-kilometer depth.

Air masses are classified according to their source region—**polar** or **tropical, maritime** or **continental**. The chief air masses that affect the weather of North America are continental polar (cP), maritime polar (mP), and maritime tropical (mT). The origin of continental polar air masses is northern Canada. In winter, the *cP* air mass is dry and stable before it moves out, but when it moves southward over the United States it is heated from below and its stability decreases. The portion that traverses the Great Lakes picks up moisture, which frequently results in snow showers along the eastern shores of the lakes and in the Appalachian Mountains. Occasionally, this *cP* air may penetrate the Rocky Mountain range.

The maritime tropical air that affects the United States east of the Rockies generally comes from the Gulf of Mexico. In winter, maritime polar air sweeping out of the Pacific is largely responsible for the winter rains of the west coast of the United States. As this air strikes the coastal range and then the Rockies, the forced lifting causes heavy rain and snow over these barriers.

After it has left its source region, an air mass can be further characterized by its temperature relative to the surface over which it is traveling. An air mass is said to be *cold* if it is colder than the underlying surface and *warm* if it is warmer than the surface. A cold air mass will be heated from below, thus the lapse rate will increase, while a warm air mass will lose heat to the underlying surface and its lapse rate will decrease (it will become more stable).

Fronts

Across a boundary separating air masses of differing properties there may exist a sharp contrast of temperature and humidity. Such a boundary, where air masses "clash," is called a *frontal zone* or, more commonly, a **front**. The name "front" was coined by the Norwegian meteorologists who first developed the polar-front theory during World War I, because the oscillations of the boundary, with periodic flareups of weather along it, reminded them of the long European battle line with its intermittent activity.

The front separating two air masses slopes upward over the cold denser air. This is illustrated in Figure 5.6, a typical vertical cross section of a cold front. Note that the vertical scale is greatly exaggerated. The average slope of fronts is only about 1:150, ranging from as little as 1:250 to as steep as 1:50. The width of the front—the

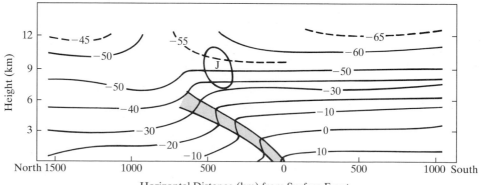

Figure 5.6 Vertical cross section from north to south through a typical cold front in middle latitudes on a winter day. Isotherms are in °C. The frontal zone is shaded and outlined by heavy lines. The position of the polar jet is indicated by J.

transition zone between air masses—is usually about 50–100 kilometers, but on the scale of distances that we are considering, such a width is closely approximated by the thickness of a heavy line drawn on a weather map.

The boundary between the warm and cold air masses must always slope upward over the cold air. This is because the cold air is denser fluid. Imagine two fluids such as water and oil, side by side, separated by a partition. If the partition is removed, the heavier water will slide beneath the oil. Now if the warm air is moving against the wedge of cold air or if the wedge is pushing under the warm air, there will be forced lifting. In either case, cooling due to expansion may lead to condensation and precipitation over the frontal surface.

Wave Cyclones

The general characteristics of the surface-weather pattern associated with migratory cyclonic depressions of the middle latitudes have been known since the late 19th century. The polar-front model associated the formation and maturation of these storms with undulations of the frontal boundary. A front separates air masses of different densities. The air masses flowing side by side may develop zones of strong wind "shear" between them; i.e., the currents of air on both sides of the boundary may have different velocities. When the flow aloft is undisturbed, fronts show little development even though temperature contrasts may be large across the front. Development of surface cyclones along fronts occurs when an upper-level disturbance approaches a front. The upper-level patterns of convergence and divergence produce surface-pressure rises and falls, respectively, and these surface-pressure changes generate low-level circulations (Figure 5.7). As the low-level cyclone intensifies, the surface front is progressively deformed as cold air is swept southeastward behind the low and warm air and moves northeastward. Successive configurations of the front resemble a developing and then a breaking wave; hence the name "wave cyclone" is frequently used to describe these systems.

Figure 5.7 Surface front and isobars and upper-level flow for developing cyclone. Upper-level divergence over the low causes pressure to fall; convergence over high produces pressure rises.

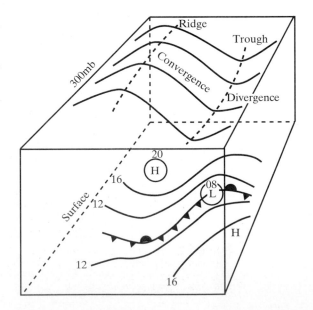

As the wave develops, low pressure forms at its apex (Figure 5.8) and both the warm and cold currents move in a cyclonic pattern around it. To the left (in the figure) of the apex, the front is advancing toward the warm air, and this segment of the front is called the **cold front**; to the right of the apex, the front is receding from the warm air and so this segment is called the **warm front**. The warm air between the fronts is know as the **warm sector**.

Figure 5.8 represents an idealized wave cyclone, the view looking downward on the Earth (shown in the center) and vertical cross sections taken a little south (bottom drawing) and a little north (top) of the apex. Imagine the entire system moving toward the right (eastward), as is normally the case. If you were standing to the east and south of the apex, ahead of the warm front, the first sign of the approaching system would be high cirrus clouds. As time goes on, the wisps of cirrus thicken to cirrostratus clouds; these often cause halos (rings around the sun or moon), a sign of rain within 24 hours, according to a well-known proverb. Gradually the clouds lower and thicken to altostratus. As the low center gets closer, the pressure falls and the wind increases and backs (changes direction in a counterclockwise direction). The temperature begins to rise slowly as the frontal transition zone approaches. Within 300 kilometers of the surface position of the front, precipitation begins, in the form of either rain or snow. After the warm front passes, the precipitation stops, the wind veers into the southwest

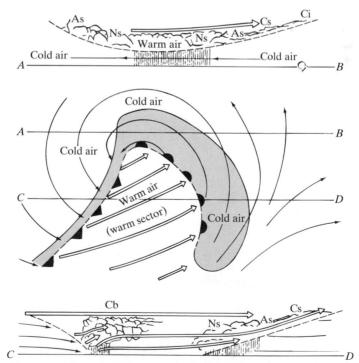

Figure 5.8   The wave cyclone model (after J. Bjerknes and H. Solberg). Center drawing, horizontal plane view; top, vertical cross-sectional view just north of wave apex (line *AB*); bottom vertical cross-sectional view across warm sector (line *CD*). (For abbreviations of cloud-type names, see section 2.1; arrows depict air flow.)

(changes direction in a clockwise direction), and the pressure stops falling. Within the warm sector, the weather depends largely on the stability of the warm air mass and the surface over which it is moving; there may be showers or almost clear skies.

The type of weather accompanying the passage of the cold front depends on the sharpness of the front, its speed, and the stability of the air being forced aloft. Often there are towering cumulus and showers along the forward edge of the front. Sometimes, especially in the midwest during the spring, several *squalls* precede the front. But in other cases, nimbostratus and rain extend over a zone of 75—100 kilometers. After the frontal passage, the wind veers sharply and the pressure begins to rise. Within a short distance behind the cold front, the weather clears, the temperature begins to fall, and the visibility greatly improves.

The genesis stage of the wave cyclone normally takes between 12 and 24 hours. Subsequent development of the wave, shown in Figure 5.9, takes an additional two or three days. As the wave breaks, the cold front begins to overtake the warm front.

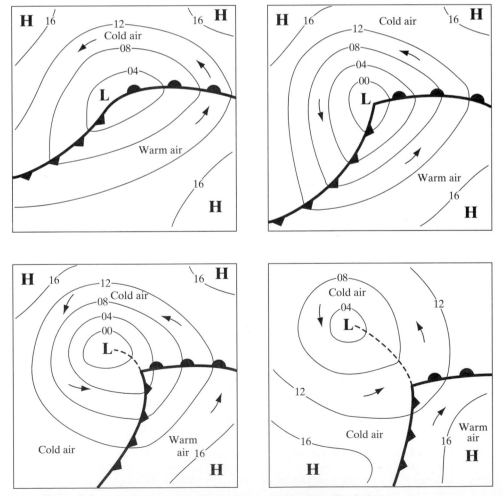

Figure 5.9 Development of a wave cyclone in the Northern Hemisphere.

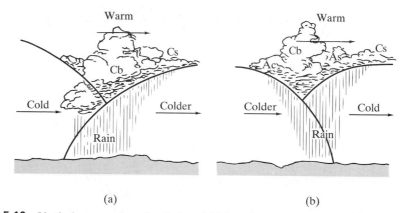

Figure 5.10 Vertical cross section of occlusions. (a) Warm-front type and (b) cold-front type. (Arrows indicate displacement of air masses.)

This process is called **occlusion** and the resulting boundary is called an *occluded front.* The vertical cross sections of Figure 5.10 illustrate that the cold front can either move up over the warm front (warm-front type occlusion) or it can force itself under the warm front (cold-front type). The occluded front is the boundary that separates the two cold air masses.

The maximum intensity of the wave cyclone, in terms of horizontal pressure gradient and wind velocity, normally occurs during the occlusion process. This happens because occlusion produces a redistribution of the air masses. The denser, cold air moves in at the surface of the system and less dense, warm air is forced aloft. This change in the distribution of mass within the eddy results in a loss of potential energy (more light air aloft, and more heavy air below) which reappears as kinetic energy—winds. Of course, the whirl is continuously being slowed by surface friction, thus losing some of its kinetic energy. When, in the last stages of the cyclone's history, there is little further readjustment of air masses and the supply of kinetic energy is cut off, friction gradually brings the giant eddy to a stop.

The net effect of the wave cyclone's history is to disrupt the initial air-mass distribution. Part of the cold air mass is swept to lower latitudes near the surface, while some of the warm air mass is transported to higher latitudes aloft.

This sequence of events associated with wave cyclones is, of course, an idealization. Few waves cyclones adhere closely to the model throughout their development. However, the model associated with wave cyclones is, of course, an idealization. Few wave cyclones adhere closely to the model throughout their development. However, the model does serve as a useful guide in furthering understanding of these vortices.

5.4 TROPICAL CYCLONES AND HURRICANES

Tropical cyclones, as their name implies, are cyclonic storms that are formed over the tropics. In fact, they almost invariably form over the oceans in the latitudes between about 5 degrees and about 20 degrees from the equator. They are spawned over all of the tropical oceans except the South Atlantic. Each area of the world has its own

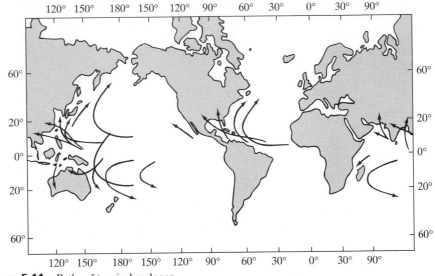

Figure 5.11 Paths of tropical cyclones.

local name for this storm, the most common being *hurricane* (North America), *typhoon* (eastern Asia), *cyclone* (India and Australia), and *baguio* (China Sea). Figure 5.11 gives the more common points of origin and paths of tropical cyclones in the world. In the discussion that follows we will use the term which is commonly used in the United States for these storms—**hurricane**.

In order to identify easily particular tropical storms in the media, they are given proper names in alphabetical order, starting over at the beginning of the alphabet at the start of each season. Different names are used for Atlantic and Pacific tropical cyclones. For example, the names of North Atlantic tropical cyclones in 1995 were Allison, Erin, Felix Humberto, Iris, Luis, Marilyn, Noel, Opal, Roxanne, and Tanya.

A satellite view of a hurricane is shown in Figure 5.12. Note how the bands of clouds spiral in a counterclockwise direction inward toward the center (Northern Hemisphere). From this photograph alone, one could hardly imagine the violence within it. Figure 5.13 presents a picture of the weather conditions in a typical hurricane. The weather that normally can be expected during the approach and passage of a hurricane can be determined by imagining the observer moving slowly (usually 15–25 km/h) from the outer edge to the center (right to left) and then out to the edge again.

High clouds, which are not too common over the tropical oceans, usually appear 300–500 kilometers in advance of the hurricane. The pressure begins to fall slowly and the winds begin to pick up above the normal 15–30 km/h of the trade winds. Within 300 km of the center, the winds reach gale force (about 50 km/h), steadily increasing in speed, and the pressure begins to fall off a little more rapidly. By the time the observer is within 160 km of the center, the winds will be 80 km/h or more, the clouds will be low and menacing, and the pressure will be falling rapidly. Rain usually starts falling 100 or 120 kilometers from the centers and increases in intensity until it is coming down in torrents at 30 or 40 kilometers from the center. Winds in the last zone may be as high as 300 km/h (185 mi/h).

Figure 5.12 Defense Meteorological satellite photograph of Hurricane David on September 3, 1979, at 10:50 P.M. E.S.T., using moonlight for illumination. (Source: Photo supplied by Henry Brandli.)

If the center of the storm passes over the observer, he has a truly startling experience. This center is known as the eye of the hurricane. Quite abruptly, the winds decrease to less than 30 km/h in a distance of 25 kilometers or less. (This distance corresponds to a time interval of less than an hour for the typical hurricane movement.) The rain stops completely, the clouds become thin, and the sun may shine through breaks. The clouds surrounding the eye appear as nearly vertical walls, extending from 1 kilometer to 12 kilometers or more. This is but a respite from the monster storm. Soon, the other half of the "doughnut" will strike and the observer will experience weather conditions similar to those encountered before, except that they will occur in reverse order and the wind direction will be opposite.

The West Indies hurricane season extends from June through November, although most hurricanes occur during August, September, and October. During the seventy-two-year period of 1887 to 1958, a total of 331 hurricanes (an average of 4.6 per

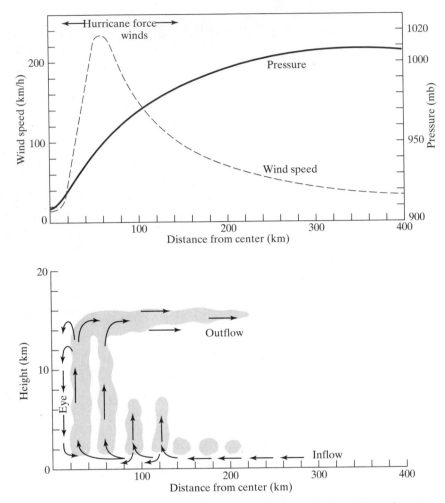

Figure 5.13 Vertical cross section through hurricane and the associated variations of pressure, wind, and precipitation with distance from the center.

year) were reported in the North Atlantic and adjoining waters, in addition to 241 other tropical cyclones that did not reach hurricane intensity (officially, having wind speeds greater than 33 m/s (74 mi/h)). About 4 percent of these occurred in the month of June, 6 percent in July, 29 percent in August, 36 percent in September, 19 percent in October, and 3 percent in November. There were only five hurricanes in the other six months of the year. There is considerable variability in the number of Atlantic hurricanes from year to year. For example, in the 40-year period from 1950 to 1990, the number of tropical storms varied from a minimum of 4 to a maximum of 14; with a mean and standard deviation of 9.3 and 3.0, respectively. During this period, the number of Atlantic hurricanes varied from 3 to 12, with a mean of 5.8 per year and a standard deviation of 2.2. Statistics have shown that the number of tropical storms is correlated with several global climatological anomalies, including rainfall in West Africa in the prior year, the direction of the winds in the stratosphere, and the El Niño phenomenon. Tropical storms are *more frequent* and *more intense* in years in which the stratospheric winds are from the *west* rather than *east* over the Equator.

There is also a relationship between the El Niño and the hurricane activity in the Atlantic. As we will see in more detail in Chapter 7, the El Niño is a phenomenon accompanied by abnormally warm ocean temperatures in the equatorial eastern Pacific. El Niño years are associated with a number of important global climate anomalies, one of them being the number of Atlantic hurricanes. Even the number of hurricanes striking the United States varies between El Niño and non El Niño years; Figure 5.14 shows the frequency of Atlantic hurricanes striking the United States varies from about one in El Niño years to two in years without an El Niño.

The average lifetime of a West Indies hurricane is nine days, although those occurring during August appear to be more durable, lasting for an average of twelve days. Hurricanes tend to move in the direction of the flow in which they are embedded, much as an eddy in a river moves downstream. During their early stages in the Atlantic, while they are still well within the easterly winds, they tend to move toward the west or northwest. If they reach north about 30 degrees latitude before dissipating, they get caught by the prevailing west winds of the middle latitudes and are swept toward the northeast.

Hurricanes sometimes move in a very erratic fashion. For example, Hurricane Flora (October 1963) meandered about over eastern Cuba for almost five days and Hurricane Betsy (September 1965) started toward the northwest over the Bahamas

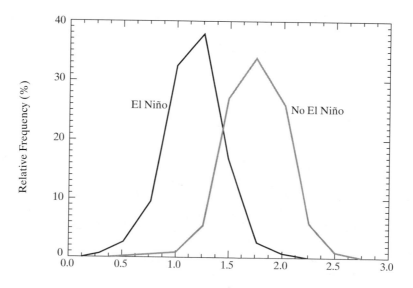

Number of Hurricanes per Year Striking the United States 1949-1992

Figure 5.14   Frequency of Atlantic hurricanes in El Niño and non-El Niño years. (Source: J. O'Brien.)

and then passed through the Florida Strait, finally crashing into Louisiana. Although the average speed of hurricanes is about 20 km/h, the speeds of individual storms are extremely variable; when they get caught up in the usual west winds north of 30 degrees latitude, they frequently greatly accelerate, sometimes achieving a speed of over 80 km/h.

The hurricane is a powerhouse of energy. Circulating hundreds of millions of tons of air at speeds of up to 300 km/h or more, the average hurricane generates 300–400 billion kilowatt-hours of energy per day, about 200 times the total electrical power produced in the United States. An average hurricane precipitates 10–20 billion tons of water each day.

The warm, moisture-laden air of the tropical oceans possesses an enormous capacity for heat energy, and most of the energy required to create and sustain a hurricane comes from what is released through condensation. A hurricane is an unusually organized, very large convection system that pumps great amounts of warm, moist air to high levels of the atmosphere at very rapid rates. The arrows in the vertical cross section of Figure 5.13 illustrate the overall convection pattern within a fully developed hurricane. Warm, moist air rises sharply in the ring between 20 and 60 kilometers of the center. New air flows in toward the center from hundreds of kilometers away. If the air starts out with even a slight counterclockwise rotary motion (caused by the Coriolis force), it will spin faster and faster as it nears the center.

The development of hurricanes occurs through the release of the latent heat of condensation in thunderstorms. Wavelike perturbations in the easterly trade winds have regions of low-level convergence associated with them, somewhat like the perturbations in the westerlies. Under especially favorable conditions, such as very warm ocean temperatures, the thunderstorms may induce a large mesoscale cyclone to form. As the air flow sin toward the center of the developing low-pressure system, additional moisture is supplied to the thunderstorms. Thus a positive feedback is established, with the developing tropical cyclone circulation supplying the thunderstorms with the required moisture and the thunderstorms providing the tropical cyclone with tremendous amounts of latent heat. The warm air produced by the latent heat release is responsible for the extremely low surface pressure. Because the tropical cyclone has a warm core, its intensity diminishes with height. Aloft, the flow becomes anticyclonic. Some of the clouds in Figure 5.12 are embedded in the clockwise outflow at the upper reaches of Hurricane David.

Hurricanes are classified by their damage potential according to a scale developed in the 1970s by Robert Simpson, a meteorologist and director of the National Hurricane Center, and Herbert Saffir, a consulting engineer in Dade County, Florida. The Saffir/Simpson scale is now in wide use and was developed by the National Weather Service to give public officials usable information on the magnitude of a storm in progress. The scale has five categories, with category 1 the least intense hurricane and category 5 the most intense. Table 5.1 shows the Saffir/Simpson scale and the corresponding criteria for classification.

The greatest damage and loss of life during hurricanes result from flooding of coastal areas by the ocean surges and waves caused by the wind. The sea is in an ag-

TABLE **5.1**   **Saffir/Simpson hurricane scale**

Category	Central Pressure (millibars)	(inches)	Winds	Surge (feet)	Damage
1	≥980	≥28.94	74–95	4–5	minimal
2	965–979	28.50–28.91	96–110	6–8	moderate
3	945–964	27.91–28.47	111–130	9–12	extensive
4	920–944	27.17–27.88	131–155	13–18	extreme
5	<920	<27.17	>155	>18	catastrophic

itated state hundreds or even thousands of kilometers from the storm center. When wind blows along a water surface it exerts a frictional drag on the water that results in ripples or "waves." The wind drag increases with higher wind speed and so does the size of the waves generated. During a hurricane, air travels at a high speed over long distances, producing waves of great height. The wave heights (vertical distance from crest to valley) often reach 10 meters and sometimes exceed 15 meters in the zone of strong winds. As waves move out from under the winds that generated them, the crests decrease in height and become more regular in shape. Waves of similar height and length between consecutive crests tend to move in groups. These composite waves are known as **swells**, and they can travel thousands of kilometers from the generating area with little loss of energy. When these swells approach a coast, the varying depth to the ocean bottom and the irregularities of the coastline complicate the wave structure. Sometimes very steep waves travel up estuaries, damaging vessels and piers. If these swells coincide with the normal high tide of the area, they may cause extensive flood damage.

However, the really damaging effects of the wind-churned ocean are not felt until the hurricane center is within a hundred kilometers or so of the coast. Rapid rises in the water level, known as **storm surges**, result from a piling up of a water along the coast by the driving winds. Such "hills" of water can be 5 meters or more above normal sea level. With storm waves riding 10 meters or more above these mounds, large inland areas can be inundated. During a hurricane in 1900, Galveston, Texas was flooded by just such a surge, which demolished the city and drowned about 5000 persons. In 1961, when Hurricane Carla struck the Gulf coast, a surge was predicted and the affected areas were evacuated beforehand, so there was no loss of human life. In 1969, Hurricane Camille lashed the Mississippi coast with winds up to 300 km/h and storm tides of 6.9 meters (22.6 ft), the highest of record. Despite the warnings, about 300 persons lost their lives as the storm moved northeastward over the middle eastern seaboard, and property damage reached $1.4 billion.

When a hurricane moves off the ocean onto land, the frictional drag that the surface of the Earth exerts on the wind is greatly increased; this slowing results in the air taking a more direct path toward the low center. This more rapid inflow toward the center leads to the gradual dissipation of "filling" of the storm, but at the same time leads to heavier precipitation.

Hurricane Andrew

On the morning of August 24, 1992, the most damaging hurricane in U.S. history, Hurricane Andrew, slammed ashore in south Florida with 145 mi/h of sustained winds and gusts exceeding 175 mi/hr. Causing $25–30 billion in property damage, Andrew the not only the most damaging hurricane on record, it was the most costly natural disaster of any kind ever affecting the United States. The devastation and effect on peoples' lives is difficult to convey with statistics; nevertheless Andrew destroyed or substantially damaged 28,000 single family homes, 9,000 mobile homes, and 11,000 housing units in multifamily structures leaving 160,000 people homeless in Dade County alone. Figure 5.15 shows some of the damage caused by Andrew. Because of the excellent forecasts and warnings and good hurricane preparedness and evacuation programs, a relatively small number (15) of deaths in Florida were directly attributed to the storm in spite of the widespread devastation. (Another 28 people in Florida died of related causes in the days following the storm.)

Figure 5.16 shows the track of Hurricane Andrew and Figure 5.17 shows the variation with time of the minimum pressure and maximum sustained wind speeds for Andrew. Unfortunately, as shown by Figure 5.17, Andrew was close to its greatest intensity when it made landfall on the 24th. The minimum pressure of 922 mb was the third lowest central pressure this century for a hurricane making landfall in the United States. The sustained winds of 125 knots created a storm tide (sum of storm surge and astronomical tide) along the coast ranging up to nearly 17 feet (Figure 5.18).

Andrew was an exceptionally well-organized storm as it made landfall in Florida. Figure 5.19 (see color insert) shows the National Weather Service radar view of Andrew at 04:35 E.D.T. on August 24, just as the eye was crossing Elliott and Sands Key. The nearly circular and symmetric eye wall (intense ring of thunderstorms) is shown by the doughnut shaped feature of maximum reflectivity. A major spiral band of thunderstorms wraps around the eye wall and is affecting much of South Florida with torrential rains and damaging winds.

Figure 5.20 (see color insert) shows a color-enhanced infrared satellite photograph of Andrew just after it crossed the Florida Coast. The nearly cloud-free eye is surrounded by an extensive circular mass of very high (and hence cold) clouds.

As shown by Figure 5.16, after crossing South Florida, Andrew went on to strike the central Louisiana coast on August 26 as a category 3 storm. Although considerably weaker, Andrew still caused 8 deaths and $1 billion of damage in Louisiana.

Trends of Property Damage and Loss of Life in Hurricanes

Figure 5.21 shows the trends in loss of life and property in the United States. The majority of the huge number of deaths shown in the first decade were due to the Galveston hurricane of 1900 in which more than 6,000 people were killed. As shown by Figure 5.21, deaths from hurricanes in the United States have been trending downward during this century, while damages have increased dramatically. Although many factors account for these different trends, a major factor in the decrease in deaths has been improved forecasts and warnings through better observations (satellites and radars) and forecasting techniques, better communications and better hurricane preparedness. The major reason for the huge increase in property loss has undoubtedly been the rapid increase in population and associated property along hurricane-

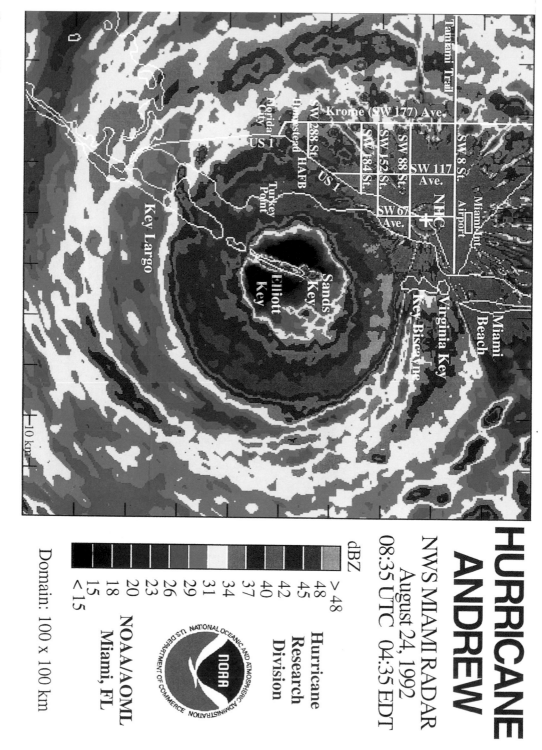

Figure 5.19 Radar photograph of Hurricane Andrew making landfall in south Florida on August 24, 1992. (Source: National Hurricane Center, NOAA.)

Figure 5.20   Infrared satellite photograph of Hurricane Andrew making landfall in south Florida on August 24, 1992. (Source: National Hurricane Center, NOAA.)

Figure 5.15 Damage in south Florida caused by Hurricane Andrew, August 24, 1992. (Photographs courtesy of National Hurricane Center.)

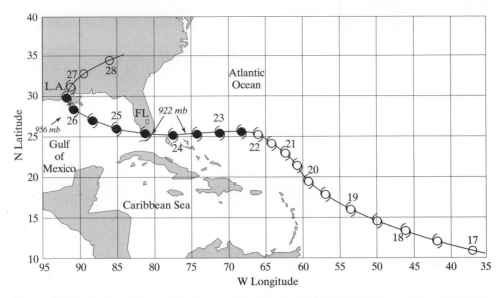

Figure 5.16 Track positions of Hurricane Andrew (August 16-28, 1992). Positions at 00 and 12 UTC* are shown. Dates are at the 00 UTC locations. Tropical depression, tropical storm, and hurricane strengths are represented by open circles and open and filled hurricane symbols, respectively.

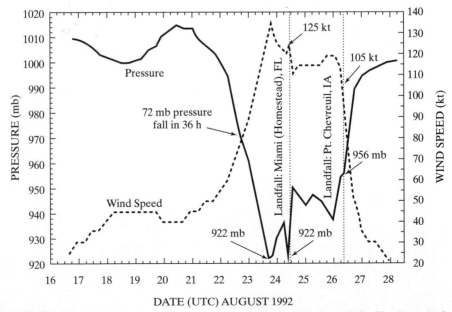

Figure 5.17 Minimum central pressures and maximum sustained wind speeds for Hurricane Andrew.

* UTC, or "Coordinated Universal Time, is the 24-hour astronomical time system and is equivalent to the local time at Greenwich, England, which is located at 0° longitude. Eastern Standard Time (EST) is 5 hours behind UTC.

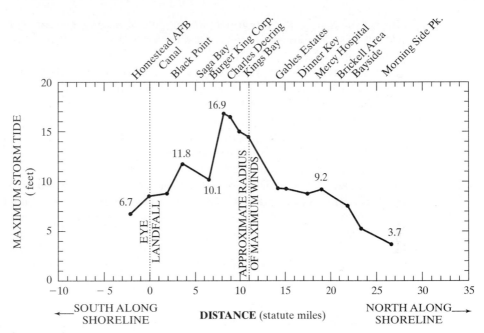

Figure 5.18 Storm tide heights (sum of storm surge and astronomical tide) along western shore of Biscayne Bay associated with Hurricane Andrew.

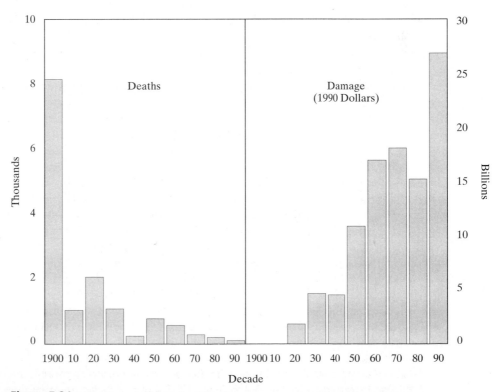

Figure 5.21 Deaths and damage from hurricanes.

prone coasts. Nearly 45 million people now live along the eastern seaboard of the United States, from Texas to Maine. Thus the people and property at risk from hurricanes is higher than ever before.

One industry that is particularly concerned about hurricanes is the insurance industry. In the wake of Hurricane Andrew, State Farm Insurance Company paid out more than $3.6 billion for losses—more than seven times the previous record. As a direct result of Andrew, as many as 11 insurance companies failed and approximately 40 companies either left Florida or have curtailed underwriting of property in Florida.

Bad as Andrew was, it could have been much worse. Studies have shown that if Andrew had tracked only 20 miles farther north in Florida, the losses would have exceeded $75 billion. Casualties would have been much higher because more than 30 percent of the region's residents did not evacuate the condominium complexes in Miami Beach, Hallandale, and Hollywood. And, as coastal populations increase rapidly, it takes longer and longer to evacuate the population; a study of southeast Florida suggests that more than 80 hours would be required to evacuate the area the next time a major hurricane threatens. Clearly the question is not whether a catastrophic hurricane greatly exceeding the death and destruction of Andrew will occur in the future; it is only a question of when.

5.5 THUNDERSTORMS

As we have already pointed out, vertical motion in the atmosphere is the key to many of the characteristics of weather. Upward motion results in expansion, cooling, and eventual condensation of the water vapor in a stream of air; the release of latent heat is often an important factor in accelerating the convection by increasing the buoyancy (instability) of the air. Downward motion results in compression, warming, and therefore an increase in the air's capacity for water vapor. We have seen, also, that convective patterns come in a large variety of sizes, the smaller ones nesting within the larger ones. In general, the maximum vertical velocity observed is inversely proportional to the size of the circulation: the large patterns have relatively feeble vertical motion while many of the small circulations have vertical motion equal to that in the horizontal.

Cloud types are closely related to the strength of the vertical motion. Stratified clouds, which sometimes extend unbroken over thousands of square miles, occur in gently ascending air (almost always less than 20 cm/s (0.4 mi/h)). Cumuliform clouds, on the other hand, occur as isolated cloud masses (individual elements rarely cover more than 75 km^2) and contain within them upward motions as strong as 30 m/s (67 mi/h). Stratified clouds form in air in which the buoyancy forces are weak or even oppose vertical motion above the thin layer in which the clouds form. For example, a wind blowing up a mountain slope may lead to condensation, but a temperature inversion may prevent vertical development of the clouds. Cumuliform clouds—those with great vertical development—are associated with instability.

The rapidity with which convective cells can develop in an unstable atmosphere is illustrated in the photographs of Figure 4.17. Of the many individual cumulus convective cells appearing on the horizon in the first photograph of Figure 4.17, one mushroomed vertically into a mature storm in only 18 minutes. It is not usual that isolated convective "cells" of this sort can be identified; normally, there is a tendency for adjacent cells to develop and join together. Frequently, there are great masses or lines of thunderstorms extending over 100 kilometers or more, but a single "cell" has a di-

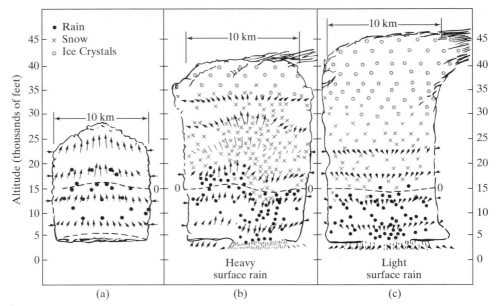

Figure 5.22 Life cycle of a typical cumulonibus cell. Schematic two-dimensional representation shows the relative wind velocities, temperature, and distribution of liquid and solid water.

ameter of about 10 kilometers.

Studies have shown that there are three characteristic stages in the life cycle of a typical thunderstorm cell. These are illustrated in Figure 5.22. The initial *cumulus* stage usually lasts for about 15 minutes. During this period, the cell grows laterally from 2 or 4 kilometers in diameter to 10 or 15 kilometers, and vertically to 8 or 10 kilometers. Note from Figure 5.22(a) that the updraft is strongest in the upper part of the cloud. Air is entering the cloud through the sides at all levels. The upward motion is actually greater than the horizontal speed, which is the reverse of what is found in larger scale atmospheric circulations.

The *mature* stage [Figure 5.22(b)] begins when rain reaches the ground and usually lasts for 15 to 30 minutes. During this stage, the drops and ice crystals in the clouds grow so large that the updrafts can no longer support them, and they begin to fall as large drops or hail. The frictional drag of the precipitation gradually slows the updraft and, in one part of the cell, a strong downward motion develops because of the precipitation and "entrainment" of drier air from outside the cloud and the resulting cooling due to evaporation of the cloud drops and precipitation. Near the center and top of the cloud, upward motion is still strong, however. Note also the strong outflow below the base of the cloud. When this downdraft meets the ground it spreads away from the thunderstorm. It is for this reason that gusty cool winds usually precede the actual arrival of a thunderstorm.

The mature stage is the most intense period of the thunderstorm. Lightning is most frequent during this period, turbulence is most severe, and hail, if present, is most often found in this stage. The cloud reaches its greatest vertical development near the end of this stage, usually reaching about 10 kilometers and sometimes penetrating the tropopause to altitudes greater than 15 kilometers.

The final or *dissipating* stage begins when the downdraft has spread over the entire cell. With the updraft cut off, the rate of precipitation eventually diminishes and so the downdrafts are also gradually subdued. Finally, the last flashes of lightning fade and the cloud begins to dissolve, perhaps persisting for a while in a stratified form.

Severe thunderstorms, which often spawn tornadoes, are considerably more complex than the more or less isolated thunderstorms of moderate intensity described previously. When the atmosphere is very unstable and abundant moisture exists in the low levels, thunderstorms may organize themselves into mesoscale circulations with typical diameters of 20 to 40 kilometers. As the thunderstorm cells pump low-level air into the upper troposphere at velocities of 40 to 80 km/h, low-level air flows in to compensate for the vertical motion. Through the conservation of angular momentum, the entire thunderstorm system may begin to rotate, producing a mesoscale cyclone. Such a circulation is shown in Figure 5.23, which shows the winds at a height of 0.3 km in a thunderstorm system in Oklahoma on June 8, 1974. The winds in Figure 5.23 were diagnosed by dual-Doppler radar measurements. Doppler radars emit a radio signal which is reflected by precipitation particles back to a receiver. The speed of the particle toward or away from the source of radiation causes a change in the frequency of the radiation (the Doppler shift). Although one radar can give only the velocity component along a line from the emitter to the particle, the three-dimensional flow can be reconstructed if there are two Doppler radars separated by some distance and observing the same particles.

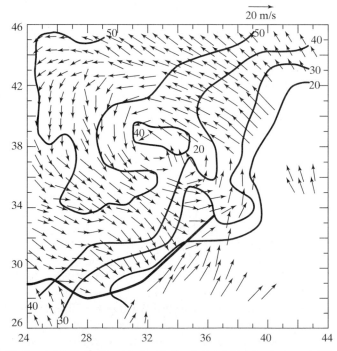

Figure 5.23   Horizontal perturbation wind analysis of severe thunderstorm near Harrah, OK on June 8, 1974. Mean flow has been subtracted. Contour lines denote relative radar reflectivity. (Source: E. Brandes, *Journal of Applied Meteorology*, April 1977.)

The winds in Figure 5.23 show strong rotation, with winds exceeding 40 m/s (144 km/h) in places. The wind shift from southeast to northwest in the southern part of the storm (indicated by a heavy black line) is the gust front and marks the leading edge of cool air originating from a downdraft.

From Doppler radar analyses such as the one presented in Figure 5.23, a reasonably complete three-dimensional picture of the circulation associated with long-lived severe thunderstorms is emerging (Figure 5.24). In Figure 5.24 a storm is moving toward the right and is being continually supplied with warm, moist, low-level air at its leading edge. In the updraft fed by this inflow, condensation produces rain below the freezing level and ice at higher levels. To the rear of the storm, dry middle-level air is incorporated into the storm. As evaporation of rain cools this air, it becomes negatively buoyant and sinks. When the resulting downdraft reaches the ground, it spreads out and forms the gust front.

Thunderstorms generally occur within moist, warm (maritime tropical) air masses that have become unstable either through surface heating or forced ascent over mountains or fronts. In the United States, practically the only source region of this air mass is the Gulf of Mexico and the Caribbean. Note how the geographic pattern of thunderstorm incidence shown in Figure 5.25 is correlated with both the distance from the source region and topography.

As Benjamin Franklin demonstrated in 1750, lightning discharges are giant electrical sparks. Cumulonimbus (thunderstorm) clouds, therefore, are great natural electrical generators that develop regions of concentrated positive and negative electricity.

Figure 4.17 illustrates how rapidly a cumulonimbus cloud can develop out of a cluster of smaller cumulus towers. Measurements have shown that small cumulus clouds are usually electrically inactive. The first indications of cloud **electrification** occur during the vigorous vertical development of large towers, accompanied by the formation of precipitation. The electric potential gradient increases rapidly and in a matter of minutes may have become high enough to cause **lightning**. The electrification process results in

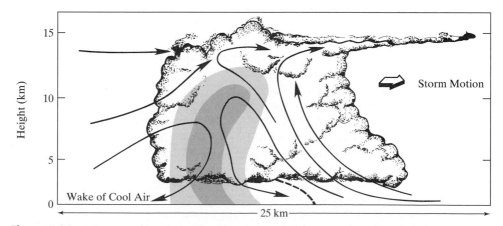

Figure 5.24 Schematic diagram showing flow in severe thunderstorm. The shaded areas represent the most intense precipitation. The arrows denote the air flow relative to the storm. The dashed line below the thunderstorm is the gust front.

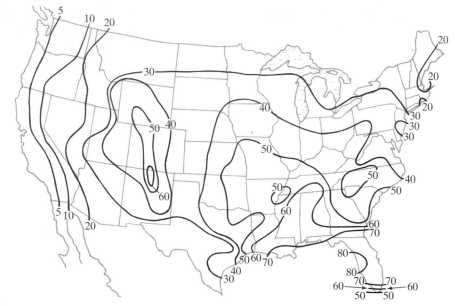

Figure 5.25 Average annual number of days with thunderstorms.

positive charges being concentrated in the upper part of the cloud with a concentration of negative charges in the lower parts. Thus, development of electrification, formation of precipitation, and cumulus cloud dynamics are closely linked physical phenomena. Researchers still strive to understand them more fully and in the meantime hold differing views concerning the basic mechanism of cloud electrification. One concept is that electrification results from an interaction between precipitation and the much smaller cloud particles. Accordingly, raindrops or hailstones make rebounding collisions with cloud droplets or ice crystals. Through a process involving friction or electrical polarization, the precipitation particles become negatively charged and the rebounding cloud particles become positively charged. The negative charge is transported downward by the precipitation and the positive charge accumulates in the upper part of the cloud. Another theory suggests that intense electrical activity results directly from the combined action of updrafts and downdrafts. Here the atmosphere, not the cloud, is considered as the source of the electric charges. The atmosphere is weakly ionized, i.e., small proportions of molecules in the atmosphere carry positive and negative charges. The concentrations of positive and negative ions are almost equal, but near the surface there is a slight excess of positive ions. Air containing this *positive space charge* enters the cloud base and is transported upward. Negative ions in the clear air are attracted toward this charge. They are intercepted at the cloud boundaries and are transported by descending air currents. As this process continues, increasing numbers of positive and negative ions are drawn into the cloud. Acting in this way, the thundercloud may be thought of as a gigantic electrostatic machine in which air currents function much like the moving parts of the machine.

Regardless of how the thundercloud does it, enormous potential differences are generated within clouds and between clouds and ground. Just before a discharge, the electrical potential gradient is of the order of 3000 volts per centimeter, and potential differences between the extremities of flashes reach hundreds of millions of volts. A typical thunderstorm generates electrical energy at an average rate of about a million kilowatts.

Special photographic techniques have shown that individual lightning discharges actually consist of multiple strokes, each lasting about 0.001 second, with about 0.04 second between successive strokes. The air along the lightning channel is heated momentarily to about 15,000 K (compared to the sun's surface temperature of about 6000 K); this causes a very rapid expansion of air, which in turn results in the sound wave called *thunder*. The rumbling of thunder occurs because sound is generated over a long discharge path, so that sound waves travel over many different paths to the observer, and much of the sound is reflected. The approximate distance from a thunderstorm can be computed by noting the time elapsed between a flash of lightning and the arrival of the sound wave and using the average speed of sound (330 m/s, or 740 mi/h).

The old proverb that lightning does not strike twice in the same place is untrue. Tall towers and buildings are repeatedly struck by lightning; Franklin's lightning rod protects such structures by providing a low-resistance conductor of the electrical current to the ground. One should always avoid being near an isolated high target during a thunderstorm. Do not get caught under a tree or in an open field during a storm.

5.6   TORNADOES AND WATERSPOUTS

The name *tornado* is probably derived from the Spanish word *tonar*, which means "to turn." A tornado is an intense cyclonic vortex in which the air spirals rapidly about a nearly vertical axis. Seen from a distance, it looks like a gray funnel or elephant's trunk extending downward from the base of a cumulonimbus cloud (Figure 5.26). Where this pendant cloud reaches the ground, great masses of dust and debris circle the lowest couple of hundred meters.

The peak winds associated with tornadoes are too strong to be withstood by the ordinary anemometer, so there are few direct measurements. Estimates from damage to buildings and the impact force of flying objects indicate that speeds range generally between 150 and 500 km/h. Such a wind necessitates a very strong pressure gradient. The largest pressure drop ever recorded with a tornado passage is 22 mb, but theoretical estimates of the maximum possible pressure drop are much higher (up to 100 millibars).

The lengths of tornado paths average only about 6 kilometers, but they are extremely erratic. Some touch ground over a distance of only 20 or 30 meters, while others hop and skip over tracks of hundreds of kilometers. Some tornadoes hardly move, while a few have been known to travel at speeds up to 200 km/h. Some last only a fraction of a minute, while others persist for several hours; the average duration is less than 10 minutes. Most move toward the east or northeast (Northern Hemisphere), but every direction of movement has been observed.

During their brief lives, tornadoes can be very destructive. A building in the path of a tornado will certainly be badly damaged, if not destroyed. The cause of the damage to buildings is two-fold: the enormous force exerted by the wind and pressure forces associated with rapidly moving air over and around building walls and roofs. Wind pressure can easily reach several hundred pounds per square foot, and powerful updrafts may lift very heavy objects. Many freak occurrences have been reported during tornadoes:

Figure 5.26   Tornado over Kansas farm. (Source: NCAR photograph.)

showers of frogs that were sucked up from ponds miles away, the "defeathering" of chickens, straws driven through posts, and entire buildings carried for hundreds of meters.

Tornadoes occur infrequently. Although they have been observed in every part of the world outside the extremely cold regions, they are most common over large continents where strong horizontal temperature contrasts exist—in the United States east of the Rockies, in the southern and middle U.S.S.R., and in southern Australia. In the United States, there are about 900 per year, mostly in the central Plains states. Iowa, Kansas, Arkansas, Oklahoma, and Mississippi have the highest frequency of tornadoes per unit area (Figure 5.27). Tornadoes occur principally in the afternoon and during the spring, but can occur at any time during the day or night throughout the year.

The formation mechanism of tornadoes is still somewhat obscure. They invariably form in association with severe thunderstorms of the type depicted in Figure

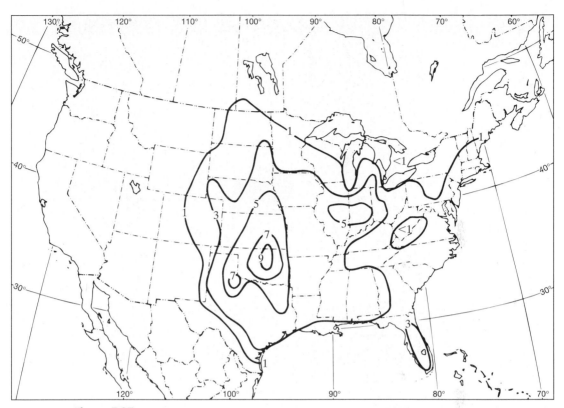

Figure 5.27 Average annual number of tornadoes per 100×100 mi. (160×160 km) square (1950–1976). (Source: *Monthly Weather Review*, 1978, p. 1175.)

5.23. The rapid updrafts in the thunderstorms produce the evacuation of air and the general lowering of surface pressures. The mesoscale cyclone (Figure 5.23) provides strong horizontal winds with considerable curvature and shear (variation of wind in the horizontal). The actual small-scale tornado funnel is possibly formed when the mesoscale cyclonic flow becomes unstable and forms smaller-scale, but more intense, vortices. Once formed, strong convection will sustain the vortex until the instability is reduced and friction destroys the whirl.

Tornado-like vortices occasionally form over warm water. Because of the high moisture content of the air, the funnels are heavily laden with water drops so that they look somewhat like a stream of water pouring from the cloud base (Figure 5.28). They are called, for this reason, **waterspouts**. Usually, waterspouts are not as intense as tornadoes over land. Near their base, the winds churn the water surface, producing waves and spray.

The **dust devil** is a small whirlwind that frequently forms on very hot days, especially over deserts. Normally, there are no clouds associated with these and they are no more than a whirling column of dust or sand. They are produced by strong convection near the surface and given a rotation by slight terrain-induced irregularities in the winds. These have been observed to rotate in both senses, clockwise and counterclockwise, with about equal frequency.

Figure 5.28   Two waterspouts over Lake Winnipeg, Canada, August 1984. (Source: Photo by Scott Norquay.)

5.7   LOCAL CIRCULATIONS

Land and Sea Breezes

A coastline is a sharp boundary between surfaces having greatly different temperature variations. The sea, because it is constantly being stirred, and because the sun's radiation penetrates the water, has a relatively small diurnal temperature change compared to the adjacent land surface. As a result, a large temperature difference can develop across the coastline during certain times of the day. In the tropics throughout the year, and at higher latitudes during the summer, the land-sea temperature gradient on a fairly calm clear afternoon can reach 15°C over a distance of less than 50 kilometers. At night, the temperature difference may be reversed (ocean warmer than land), although normally not nearly so pronounced.

These land-sea temperature differences lead to the creation of a thermal circulation such as that shown in Figure 5.29. The daytime landward flow is known as a **sea breeze**, while the seaward flow at night is called a **land breeze**. Like the monsoon, the sea breeze can cause rainfall, although it frequently only induces the cloudiness if the lifting is insufficient to cause precipitation. In the tropics it may occur all year round, but at higher latitudes it is mostly a summer phenomenon.

The sea breeze usually begins to develop three or four hours after sunrise, because air over the land heats up more rapidly than air over water. By 1 or 2 P.M., when it has reached its peak intensity, the circulation cell usually extends both inland and sea-

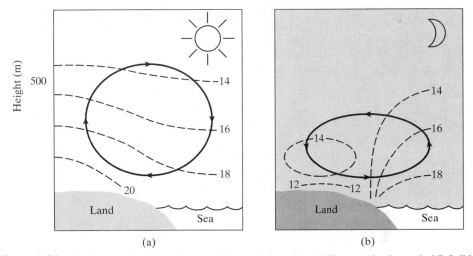

Figure 5.29 Sea-breeze (a) and land-breeze (b) circulations. Dashed lines are isotherms in °C. Solid lines indicate direction of flow.

ward about 20 kilometers, although it has been found to penetrate inland as much as 60 to 70 kilometers. The entire circulation cell, including the upper, seaward flow, is not normally more than 1 kilometer deep, although in the tropics it may reach 3 kilometers or more. As the forward edge of the sea breeze passes over a point in its landward penetration, the relative humidity increases and the temperature decreases sharply (the former by 40 percent or more and the latter 5°C or more in less than an hour). Sometimes fog or low stratus clouds may accompany the sea breeze; there are places where the coastal water is extremely cold, such as along the Peruvian coast, where the forward edge of the sea breeze is so sharply defined that the fog appears as a solid wall.

As the land cools in the evening, the sea breeze dies, and between about 7 and 10 P.M., there is little evidence of it. The land breeze, much weaker than its daytime counterpart, will normally begin at 10 or 11 P.M. and reach its maximum development near sunrise. The principal effect of the land breeze is to prevent the air over the land from cooling quite as much as it otherwise might.

The sea breeze plays an important role in moderating the temperature of narrow strips of land along seacoasts and lake fronts. The breeze created by lakes is generally much less intense and has a smaller width and depth. Along the shores of the Great Lakes, for example, the inland penetration is usually not more than a few kilometers. However, it offers a welcome relief from the summer heat for residents who live close to the shore.

Mountain and Valley Winds

Along mountain slopes, a thermal circulation occurs that also has a diurnal cycle (Figure 5.30). During the daytime, from about three hours after sunrise until sunset, an upslope wind, called a **valley wind**, blows. Between about midnight and sunrise, an opposite, downslope wind, called a **mountain wind** occurs. The mountain-valley winds are most pronounced on clear summer days, when the prevailing winds are weak.

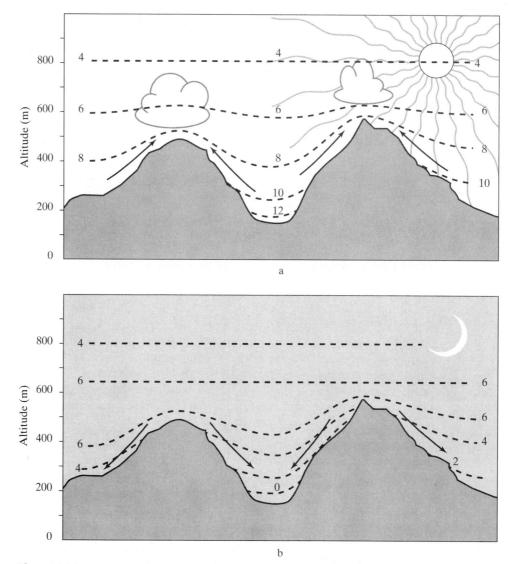

Figure 5.30 Mountain breeze (a) and valley breeze (b) circulations. Dashed lines are isotherms in °C. Solid lines indicate direction of flow.

The mountain-valley circulation is produced because the air in contact with the slope is either warmer and less dense (daytime) or cooler and denser (nighttime) than air in the same elevation over the valley. As a result, the air over the slopes rises during the day and sinks at night. Of course, the intensity of the flow and the specific direction at any point depend on the degree of slope and the configuration of the valley. Mountain and valley winds are best developed in wide, deep valleys.

The rising air currents along mountain slopes are a familiar phenomenon to every mountain climber. Frequently, cumulus clouds and showers form over summits in the ascending, expanding air. The depth of these rising currents above the slopes is usually between 100 and 200 meters.

Katabatic Winds

All downslope, drainage-type winds are referred to as **katabatic winds**. Most are weak, usually not exceeding 10 km/h, (6 mi/h), and are significant primarily because they cause cold air to drain into the valley, producing lower night temperatures in the valley than on the mountainside.

There are some very strong drainage (katabatic) winds, but most of these are set into motion by the large-scale or prevailing flow. One of the strongest of these, the glacier wind, may attain destructive violence. It occurs when air is cooled as it moves across snow fields on high plateaus; at the edge of the plateau the air cascades downward. At some places, such as along the fjorded coasts of Norway, Greenland, and Alaska, deep canyons channel the flow, thus augmenting the speed considerably. These winds blow during both the day and night. Another example is the **bora**, which sporadically brings in cold air down rather steep slopes to the usually warm Adriatic Sea. The bora is an intermittent wind, gusts of 50 to 100 km/h being interspersed with calms. Where it reaches the sea, the bora produces great waves and kicks up sprays in great quantities. A similar cold wind, the **mistral**, occurs along the French Mediterranean coast.

Föhn Winds

The **föhn** is a downslope wind that occurs in many mountainous areas, but it is not caused by drainage of dense air. It is a warm, very dry, erratic wind that sometimes appears along the lee slopes of mountain ridges. It occurs when the prevailing winds in humid air are directed against a mountain. The forced ascent causes thick clouds to form and, on occasion, heavy orographic precipitation. During most of the ascent, cooling proceeds at the wet adiabatic rate (approximately 6°C/km) and by the time the air reaches the peak level, much of its moisture has been removed. After crossing the ridge line, some of the air descends along the lee slopes, warming the *dry* adiabatic rate (10°C per kilometer). When it arrives near the bottom of the mountain, the air is very warm and dry, having been heated by adiabatic compression.

Although föhn winds are observed along many mountain ranges in the world, some of the most extreme cases occur along the eastern slopes of the Rocky Mountains. Here the phenomenon usually goes by the name of **chinook**, the Indian territory from which they seem to come. The Indians commonly referred to this wind as the "snow eater" because its extreme dryness and warmth could melt and evaporate as much as a meter of snow in a day. The chinook frequently forces out the cold air that lies along the eastern slopes, and the temperature can rise extremely rapidly when the chinook arrives. On the morning of January 22, 1943, the temperature at Spearfish, South Dakota, rose from –4°F at 7:30 A.M. to 45°F at 7:32 A.M.—a rise of 49°F in two minutes.

KEY TERMS

air mass
angular momentum
anticyclone
bora
centripetal force
chinook
cold front
continental
cyclone
doldrums
dust devil
electrical charges
electrification
Föhn
front
general circulation
horse latitudes

hurricane
jet stream
katabatic wind
land breeze
lightning
maritime
mesoscale
microscale
mistral
monsoon
mountain wind
occlusion
polar
polar easterlies
polar front
prevailing westerlies

scales of motion
sea breeze
storm surge
swells
synoptic scale
thunderstorm
tornado
trade winds
tropical
tropical cyclone
valley wind
vortices
warm front
warm sector
waterspout
wave cyclone

PROBLEMS

1. Compute the kinetic energy per unit mass, $V^2/2$, for each of the following vortices:
   a. Tornado: Radius = 0.2 km, speed = 250 km/h, height = 2 km
   b. Hurricane: Radius = 50 km, speed = 150 km/h, height = 8 km
   c. Extratropical cyclone: Radius = 500 km, speed = 50 km/h, height = 8 km
   Estimate the total volume of air circulating around each vortex, and from the mean density in the vertical (Appendix 2), calculate the total mass in each circulation system. Compute the total kinetic energy ($mV^2/2$) in kilowatt-hours for each system. Compare the results with the electrical energy production in your city.

2. List the energy sources of each of the major atmospheric vortices. Discuss the theories of how each vortex is initiated. Very few vortices are known to occur over polar regions in winter. Why?

3. During what season and in which hemisphere would you expect the sharpest transition zones (frontal boundaries) between air masses? Why?

4. In which sense does the vortex turn over your bathtub drain? Is the Earth's rotation responsible? (Consider the magnitudes of the forces.) Would you expect any rotation if the bathtub were on the equator?

5. Considering the surface winds of the general circulation, by what general route would sailing ships travel from London to New York? By what route would they return? Which of these routes would be the shorter distance?

6. Since the northeast trade winds prevail over most of the Northern Hemisphere between latitudes 10°N and 25°N, why are the winds over southeast Asia from the south during the summer?

7. In a given location, why is the sea breeze usually stronger than the associated land breeze?

8. On the average, the annual temperature range in the Northern Hemisphere is much greater than that in the Southern Hemisphere. Why?

9. An examination of the normal sea-level pressure distribution for July shows a significant low-pressure system located over the area of the Mojave Desert, yet almost no precipitation falls in that region during the summer. Why?

6
Weather Forecasting

Accurate prediction is a major goal of science. But few physical scientists have a more complex, more frustrating, or more challenging medium to work with than the meteorologist. It has been pointed out in previous chapters that circulations of sizes ranging from millimeters to thousands of kilometers—a range of nine orders of magnitude—exist in the atmosphere, that the Earth's surface is not only rough and mountainous but is covered with different materials, and that even the constituents of the air, especially that the very important one, water vapor, vary considerably in both space and time. Such intricate and ever-changing weather patterns can be only very inadequately observed and it is little wonder that improvements in the accuracy of prediction have been slow. Nevertheless, as discussed in this chapter, there have been improvements—improvements made possible by better observations, greater understanding, and computer models of the atmosphere.

6.1 WEATHER MAP ANALYSIS

Weather forecasting starts with atmospheric observations. More than 10,000 land stations and hundreds of ships take regular observations at the surface, and more than 1000 stations make regular radiosonde observations at least once a day over the world. In addition, commercial aircraft take observations over land and oceans along their flight paths.

Since the first meteorological satellite, TIROS-1, was launched on April 1, 1960, satellites have played an important role in providing first meteorological images, and later temperature, moisture, and wind date for use in weather prediction. Atmospheric soundings are now provided by satellites in polar and geostationary orbits. The United States operates two Geostationary Operational Environmental Satellites (GOES) located over the Equator. They provide frequent cloud images, from which cloud motions are measured to estimate winds at cloud altitudes; a GOES image for June 21, 1995 is shown in Figure 3.6.

Through international agreement, all nations of the world exchange their meteorological data, except in time of war. To facilitate communication, all nations use a standard weather-reporting code.

After the forecast offices of each country receive the data, the first stop is to prepare a three-dimensional analysis of the atmosphere. There are many techniques for studying the variation of atmospheric properties in both the horizontal and vertical, but the most widely used method is the construction of a series of charts that represent horizontal cuts of the atmosphere from sea level to above the tropopause. The **surface map** is far by the most complete, both in terms of the number of stations reporting and the number of variables represented at each point. This is the familiar weather map that appears in many newspapers and on television. An example of a surface-weather map is given in Figure 6.1. At the location of each reporting site, its complete observation is plotted, following the model shown in Appendix 3. A vast amount of information is packed into the surface data; a complete description of the clouds, temperature, humidity, pressure, wind, precipitation, and restrictions to visibility.

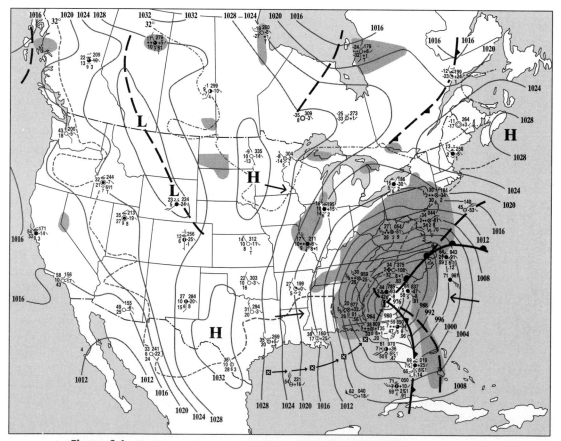

Figure 6.1 Surface weather map for 7:00 A.M. E.S.T., March 13, 1993.

Conditions aloft are usually represented by charts of **isobaric** (constant-pressure) **surfaces**. These are equivalent to charts of constant altitude, since the horizontal pressure gradient at a constant level is proportional to the slope of an isobaric surface (see Figure 4.2). On constant-pressure charts, the wind conforms to the gradient of height contours in the same way that the wind is represented by isobars on a constant-altitude chart. The plotting model used for these "upper-air" charts is given in Appendix 4.

Examples of analyzed upper-air charts (500 millibars, approximately 5500 meters) are given in Figures 4.4 and 4.5. The constant-pressure surfaces that are routinely analyzed are 850, 700, 500, 300, 200, and 100 millibars (approximate altitudes of 1500, 3000, 5500, 9200, 11,800, and 16,200 meters).

Many other types of analyses are regularly made at most meteorological forecast centers—vertical cross sections of the atmosphere showing the distribution of temperatures, winds, and moisture in areas of particular concern; analyses of wind speed that help to locate the jet stream; analyses of atmospheric stability for the prediction of thunderstorms and tornadoes; and others that describe the properties of motion (e.g., the air's "spin"), potential energy, etc.

6.2 FORECASTING TECHNIQUES

After the current state of the atmosphere has been determined by analyses such as those described in the previous section, the next difficult step is to determine the future patterns of the meteorological variables (temperature, wind, etc.). Several techniques have been employed in short-range forecasting, but only the two most widely used methods will be discussed here. The first of these involves the use of rules and formulae to determine the displacement and changes in intensity of such weather-map features as lows, highs, fronts, waves aloft, and the jet stream. Because this method directs its efforts at specific prominent features of the weather map rather than taking into account the complete field of temperature, moisture, and momentum, the accuracy that can be achieved is limited. This technique depends greatly on highly idealized models of atmospheric phenomena, such as that of the wave cyclone discussed in Chapter 5. But atmospheric features such as fronts and cyclones do not conserve their properties with time; they are affected strongly by their environment and change in size and intensity, as well as location.

The patterns of flow represented on the upper-air charts [as illustrated by Figure 4.4] do have certain characteristics of behavior that are helpful to the forecaster. The sea-level chart is generally dominated by several closed isobaric systems—cyclones and anticyclones—but aloft the picture changes considerably. There are fewer closed systems in the middle and upper troposphere; rather, a general westerly flow undulates around the poles. In middle latitudes, between 40 degrees and 50 degrees, these westerlies usually attain a peak velocity. Four or five major waves in the westerly-flow pattern normally can be identified around a hemisphere, with many minor oscillations superimposed. These small undulations are associated with the fast-moving, short-lived (3 or 4 days) wave cyclones near the surface, while the longer waves are associated with the larger scale, more sluggish features of the weather. These waves are closely associated with the distribution of clouds and precipitation. Ahead of upper-level troughs the air is generally rising, while behind the troughs the air sinks, warms and dries. Thus, when an area is located to the west of the trough of one of these long waves, there is likely to be a 2- or 3-week period of dry weather, while if it is to the east of the trough, wet weather is likely to prevail for a few weeks.

Prediction from Equations

Meteorologists have long dreamed of being able to compute the future state of the atmosphere, much as the astronomer computes a future eclipse. The basic physical equations governing the behavior of fluids have been known for almost a century, and they have been applied to the atmosphere for more than 50 years. In principle, meteorologists should be able to solve these equations to provide weather forecasts. But early attempts to do this ended in failure, and only since 1949 have practical, although as yet imperfect, computer forecasts been made on a regular basis.

The basic principle of large-scale numerical weather prediction is to write predictive equations for the variables that represent the weather, i.e., horizontal wind components, temperature, surface pressure, and water-vapor content. For example, the equation for the rate of change of temperature with time at a fixed point in space is

$$\frac{\Delta T}{\Delta t} = (\text{advection} + \text{pressure change} + \text{diabatic}) \text{ effects} \qquad \textbf{(6.1)}$$

A mathematical version of equation (6.1) involving the other atmospheric variables are written to represent the rate of temperature change due to advection (horizontal and vertical transport of air of different temperatures), changes in pressure (expansion or compression), and diabatic effects (heating or cooling). Equations similar to (6.1) for the winds and water vapor are written for many points at different locations and levels in the atmosphere. For example, a domain the size of North America and adjacent oceans might be covered by a 50×50 horizontal **grid** at 10 levels, representing 25,000 separate points. Given an **initial state** of the atmospheric variables at all of these points, the time rate of change may be computed from the prognostic equations, which then yield the change of each variable at each point over a small time interval (say 10 minutes). From this slightly advanced (in time) data set, the process may be repeated, yielding even later estimates of the variables. By repeating this cycle hundreds of times, the horizontal and vertical distribution of the atmospheric variables can be predicted for the future. All of these calculations are performed on high-speed computers in a matter of minutes.

Although the above procedure could theoretically be carried out for months and years into the future, there are a number of serious problems that cause the forecasts to deteriorate with time. First, the mathematical equations that describe fluid motion are non-linear—a type that cannot be solved simple; numerical techniques, which are laborious, requiring an immense number of computations, must be used. Until the invention of high-speed computers, it was impossible to do these in a reasonable time. Second, observations exist in insufficient detail over much of the Earth to permit an accurate representation of the fluid motion everywhere at the initial time. Gaps in the data and errors in the observations introduce errors that gradually increase with time, because no portion of the atmosphere is independent of the rest. (In other words, the weather 1000 miles away may provide the "seed" for tomorrow's weather.) Finally, there are difficulties in representing accurately the physical processes in the atmosphere, such as how to take into account the highly variable effects of friction, mountain barriers, and the distribution of heat and cold sources.

By making certain simplifications in the equations, the world's most powerful computers now produce forecasts of the fields of horizontal and vertical motion on a routine daily basis. Despite the simplifying assumptions, the results are much better than those produced by the older, subjective methods. Although inherent limits to atmospheric predictability prevent skillful daily weather forecasts for more than about 2 weeks in advance, numerical prediction is a major stop ahead in the science.

Significant progress in weather prediction has been made over the past 40 years. Figure 6.2 shows the steady improvement in the skill of predicting the 500-mb height field over the United States by the National Centers for Environmental Prediction, which is located in Washington, D.C. This improvement is due to a variety of factors including development of better computer models and improved ways of analyzing atmospheric data. Similar improvements have occurred in the prediction of weather patterns at other levels, including the surface.

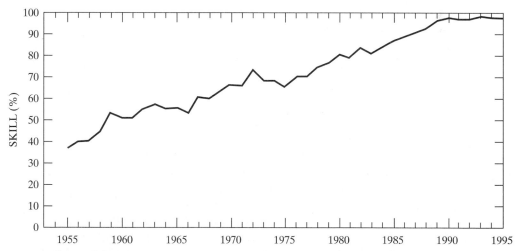

Figure 6.2   History of operational 36-hour forecasts of the heights of the 500-mb constant pressure surface over the United States by the National Centers for Environmental Prediction (NCEP), National Weather Service, NOAA in Washington, D.C. A skill of 100% means that the pattern of 500-mb contours, which is closely related to the circulation at that level (approximately 18,000 feet above sea level) is essentially forecast perfectly. Because many other factors determine the weather at the surface, 36-hour forecasts of weather phenomena such as clouds and rain are still not perfect at 36 hours even though the 500-mb height forecasts at 36 hours contain only very small errors on the average. (Source: E. Kalnay and M. Pecnick, NCEP.)

Weather Forecasting

The extrapolation of prominent characteristics in the weather pattern is still the most generally used forecasting method for short (0–6 hours) time periods. For longer time periods, computer model forecasts become more important. After the forecaster has decided on what the future weather maps will look like—where the high and low centers, fronts, etc., will be located and how intense they will be—he is still faced with the problem of relating the forecast pattern to the minutiae of "weather." Exactly where will there be precipitation and what type? Where will the clouds be and what will be the temperature over New York? To forecast these things, the meteorologist considers the factors that are likely to produce modifications in the idealized models that he has employed. For example, he must estimate whether the heating by a surface of the forced ascent of flow over a hill will be sufficient to release instability that may cause showers in an otherwise clear air mass.

The reliability of weather forecasts decreases markedly as the time interval over which they are projected increases, and so it is customary to distinguish among **short-range** (less than 48 hours), **extended-range** (up to about a week), and **long-range** (a month or longer*) forecasts. The last two are often referred to as *outlooks* and are usually quite general in character. The long-range forecasts merely state that the weather is expected to be colder or warmer, more rainy or less rainy than normal for the area and time of year.

Forecasting the March 1993 Superstorm

During the period of March 12–14, 1993, one of the most intense extratropical cyclones to develop over land in years paralyzed much of the eastern United States. Called by many "the Storm of the Century," this cyclone brought heavy snow and blizzard conditions, tornadoes, coastal flooding, and record cold to the eastern half of the United States, Cuba, and parts of northern Mexico. The storm resulted in close to a hundred fatalities, hundreds of injuries, damage that exceeded $2 billion, and the most widespread disruption of air travel in the history of aviation. Fortunately, this storm was well forecast, with early warnings from the operation global weather prediction models that a major winter storm would affect the East Coast over the weekend being provided as early as Monday of the previous week.

A history of the storm is provided in Figure 6.3. The areal distribution of significant snowfall is among the most widespread of any storm ever recorded in the eastern United States; for example, the area enclosed by the 10-inch (25 cm) contour exceeds 300 percent of the area covered by the infamous blizzard of March 1888. Record-breaking snowfall totals included 13.0 in (33 cm) at Birmingham, Alabama; 20.0 in (51 cm) at Chattanooga, Tennessee; 30.9 in (78 cm) at Beckley, West Virginia; and 42.9 in (109 cm) at Syracuse, New York.

---

*Of course, publishers of almanacs and calendars do not hesitate to forecast a year of more in advance, and there are a few private weather forecasters who will undertake to forecast the weather many months away. However, their accuracy is generally not above that obtained from "climatology" (i.e., from predicting random variations of the average or normal weather); if they have some scientific method, it is a well-guarded secret.

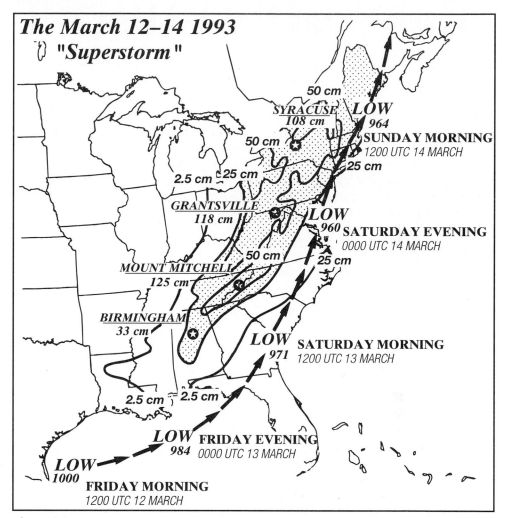

Figure 6.3 "Storm of the century" track with 12-h positions, central sea level pressure (mb), and total snow fall (cm). (Source: Figures 6.3, 6.4, 6.5, 6.7, and Table 6.1 were published in the February 1995 issue of the *Bulletin of the American Meteorological Society* and were provided by Paul Kocin, NOAA.)

The superstorm also caused flooding in Florida and Cuba and tornadoes in northeastern Mexico and Florida. The flooding in Cuba caused damage of $1 billion; in Florida at least 47 lives were lost and damages exceeded $1.6 billion. The storm caused numerous record low sea-level pressures and record low temperatures (See Table 6.1).

The meteorological evolution of the storm is illustrated in the surface and 500-mb weather maps for 00 UTC Saturday 13 March (7 P.M. E.S.T., Friday, March 12) through 1200 UTC March 14 (7 A.M. E.S.T. Sunday, March 14) in Figures 6.4 and 6.5. On Friday evening, an already intense cyclone is located over the Gulf of Mexico (Figure 6.4(a)). Heavy rain is occurring along the Gulf coast, with snow beginning

TABLE 6.1   List of record-low sea level pressures (inches, kPa), temperatures (°F; °C), and selected maximum wind gusts for the United States (mph; m s⁻¹). All sea level pressures were measured on 13 March. All temperatures occurred on March 14–15 and all wind reports occurred on March 13 (except for South Timbalier and South Marsh Island, Louisiana, which occurred on March 12).

Record-low sea level pressures	Inches	kPa	Record-low temperatures	°F	°C
White Plains, NY	28.38	961.1	Burlington, VT	-12°	-24.4°
Philadelphia, PA	28.43	962.4	Caribou, ME	-12°	-24.4°
New York, (JFK), NY	28.43	962.4	Syracuse, NY	-11°	-23.9°
Dover, DE	28.45	963.0	Mount LeConte, TN	-10°	-23.3°
Boston MA	28.51	965.0	Elkins, WV	-5°	-20.6°
Augusta, ME	28.53	965.7	Waynesville, NC	-4°	-20.0°
Norfolk, VA	28.54	966.0	Rochester, NY	-4°	-20.0°
Washington, DC	28.54	966.0	Pittsburgh, PA	1°	-17.2°
Raleigh, NC	28.61	968.6	Beckley, WV	1°	-17.2°
Columbia, SC	28.63	969.3	Asheville, NC	2°	-16.7°
Augusta, GA	28.73	973.0	Birmingham, AL	2°	-16.7°
Greenville-Spartanburg, SC	28.74	973.4	Knoxville, TN	6°	-14.4°
Asheville, NC	28.89	978.4	Greensboro, NC	8°	-13.3°
			Chattanooga, TN	11°	-11.6°
			Philadelphia, PA	11°	-11.6°
			New York, NY	14°	-10.0°
			Washington, DC	15°	-9.4°
			Montgomery, AL	17°	-8.3°
			Columbia, SC	18°	-7.8°
			Atlanta, GA	18°	-7.8°
			Augusta, GA	19°	-7.2°
			Mobile, AL	21°	-6.1°
			Savannah, GA	25°	-3.9°
			Pensacola, FL	25°	-3.9°
			Daytona Beach, FL	31°	-0.6°

Highest recorded wind gusts	mph	(m s⁻¹)
Mount Washington, NH	144	(64.4)
Dry Tortugas, FL	109	(48.7)
Flattop Mountain, NC	101	(45.2)
South Timbalier, LA	98	(43.8)
South Marsh Island, LA	92	(41.1)
Myrtle Beach, SC	90	(40.2)
Fire Island, NY	89	(39.8)
Vero Beach, FL	83	(37.1)
Boston, MA	81	(36.2)
LaGuardia Airport, NY	71	(31.7)

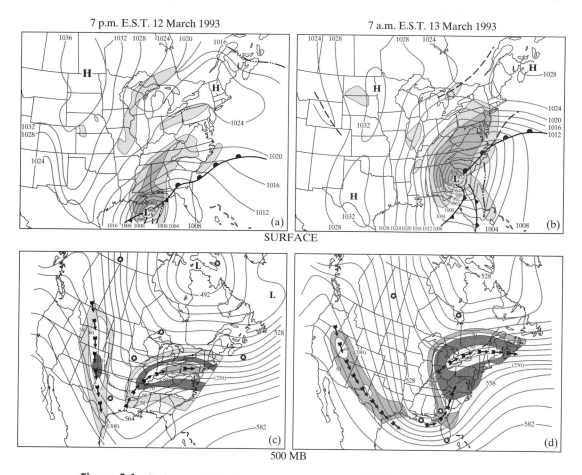

Figure 6.4 Surface and 500-mb weather maps for 7 p.m. E.S.T March 12 and 7 a.m. E.S.T. , March 13, 1993. Shading on surface maps denotes areas of precipitation. Shading on 500-mb denotes regions of maximum wind. Isotachs are labeled in m/s height contours in decameters.

over the southern Appalachians. The low-level wind flow over most of the eastern United States is from the north or northeast, a result of the strengthening pressure gradient between the developing cyclone and the two strong anticyclones, one centered over the Dakotas and the other over New England.

At 500 mb (Figure 6.4(c)) a sharp trough of low pressure extends from the intense upper-level low over Hudson Bay southward across the central United States. Associated with the strong pressure gradient (as revealed by the close spacing of the 500-mb height contours in Fig. 6.4(c)), are two regions of maximum winds, one to the west of the trough axis over New Mexico and Colorado and one to the east, over the middle Atlantic states. These relatively localized areas of maximum winds embedded in the overall jet stream are called **jet streaks**, and they are usually associated with active weather and storms. In particular, when a strong jet streak occurs to the rear (west) of an upper-level trough and surface cyclone, as is the case here, the trough and the cyclone usually intensify rapidly.

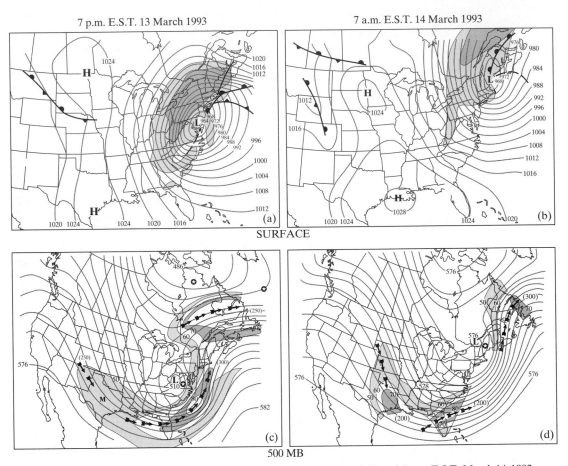

7 p.m. E.S.T. 13 March 1993   7 a.m. E.S.T. 14 March 1993

SURFACE

500 MB

Figure 6.5 Surface and 500-mb maps of 7 p.m. E.S.T March 13 and 7 a.m. E.S.T., March 14, 1993.

During the night, the cyclone moved northeastward and intensified, crossing the northwest coast of Florida just west of Tallahassee around 3 A.M. E.S.T. A squall line, accompanied by numerous severe thunderstorms and tornadoes moved across Florida during the night. Created by gale-force winds from the west, coastal flooding reached a depth of 10–12 feet (3–4 m) along the northwestern coast of Florida.

The situation at 7 A.M. E.S.T. on Saturday morning, is depicted in Figure 6.4(b) and 6.4(d). At the surface (Fig. 6.4(b)), the storm is now centered over central Georgia, with the minimum pressure dropping from 984 mb to 971 mb overnight. The pressure gradient has also intensified, and with it the surface winds around the cyclone. The persistence of the strong pressure gradient and associated westerly gales maintained high water across western Florida for many hours, including long after the cold front passed.

At 500 mb (Figure 6.4(d)) the trough has intensified as predicted and cold air aloft (represented by the low heights of the 500-mb surface) extends to the Gulf coast. A very strong jet stream occurs around the base of the trough, and embedded in the jet stream are two strong jet streaks, one centered over the southwestern United States and one over the northeastern United States. The jet streak to the rear of the trough indicates that there is still potential for deepening of the trough and surface cyclone.

Twelve hours later, on Saturday evening, the storm has intensified further and is now centered over Delaware (Figure 6.5(a)) with a minimum pressure of 960 mb (lower than many hurricanes). The storm is at its maximum intensity at this time, with heavy snows and hurricane force winds and wind gusts occurring over much of the eastern United States. At 500 mb (Figure 6.5(c)) the trough has deepened further, and, in fact, a closed low with central heights below 5100 meters (indicated by the 510 contour on Figure 6.5(c) has formed over West Virginia. The highest winds around the trough (greater than 80 m/s or 160 mph) are now located ahead of and to the east of the upper-level trough, indicating that further intensification of the storm is unlikely.

A visible satellite photograph of the eastern United States taken at 10:30 A.M. E.S.T. on Saturday morning (Figure 6.6) gives a good view of the areal extent of the storm and its intensity. At this time the center of the storm is located near the border between South Carolina and North Carolina, very near the northwestern tip of the wedge of nearly clear air that has been drawn in toward the storm center. The long, thick band of clouds extending from the low center southeastward off the coast of the United States represents heavy showers and thunderstorms associated with the cold front. The thick, nearly unbroken cloud mass extending from northeastern Florida, Georgia, and Alabama northward into southern Canada is producing the heavy snow

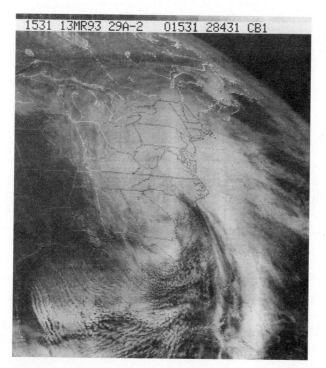

1531 13MR93 29A-2   01531 28431 CB1

Figure 6.6   GOES-7 visible satellite photograph of 10:30 E.S.T., March 13, 1993. (Source: P. Kocin, NOAA.)

and rain falling over much of this area. The broken clouds over the Gulf of Mexico, which sweep around the cyclone over Florida and over the Atlantic, are low cumulus and stratocumulus clouds formed as the extremely cold dry air in the northerly and westerly flow around the cyclone flows over the relatively warm water and is heated and moistened from below. This produces unstable conditions and the extensive formation of convective clouds.

The final maps of this sequence (Figure 6.5(b) and 6.5(d)) depict the storm on Sunday morning. By this time, the center of the storm has moved to the Maine coast and the central pressure has risen a bit, from 960 to 964 mb, indicating that the storm is slowly weakening. The storm's effects are far from over, however, as near hurricane-force northwest winds rake most of the eastern United States to the rear of the storm's center. At 500 mb (Figure 6.5(d), the closed upper-level center of low pressure is located nearly over the surface low. Cold air has swept nearly completely around the storm, indicating that the storm has reached the *occlusion* stage. With the weakening of the horizontal temperature contrasts in the vicinity of the cyclone center, the storm's source of energy is diminished and the storm will begin to weaken slowly.

By Sunday afternoon, the storm had left the United States and snowfall was gradually ending over New England. Clearing skies over the eastern United States allowed for further radiational cooling of the already frigid air mass and set the stage for the record low temperatures over much of the region (See Table 6.1)

The forecasts of this case of extreme cyclogenesis, while not perfect, illustrate the tremendous progress made over the past 40 years in weather forecasting, in which forecasting evolved from largely an art to mostly a science. The cyclogenesis was forecast by the global computer models 5 to 6 days in advance, long before any surface data indicated the formation of a storm. The unusual intensity of the storm was forecast three days in advance, allowing forecasters, government officials, and the media ample time to warn the public, marine, and aviation interests to take precautions for the protection of life and property. The excessive amounts and areal distribution of snowfall were predicted two days before the first snowflake fell. Blizzard watches and warnings were issued throughout the eastern United States with unprecedented lead times (Figure 6.7). These outstanding forecasts and warnings undoubtedly prevented much loss of life, injuries, property damage, and public inconvenience.

However, the forecasts were not perfect, especially during the early stages of the storm when the models failed to predict the degree of intensification over the Gulf of Mexico. These model errors led to errors in forecasting the intensity of severe weather and the impacts over Florida, in particular. In spite of these errors, however, the excellent forecasts overall indicate the enormous progress that has been made in the science of weather prediction over the years.

Extended and Long-Range Forecasts

Because of the growth of errors in the initial conditions of numerical weather prediction models, and errors in the representation of physical processes in the models themselves, the models generally show skill in **deterministic** weather prediction (the prediction of specific weather events such as "the Storm of the Century") for only a

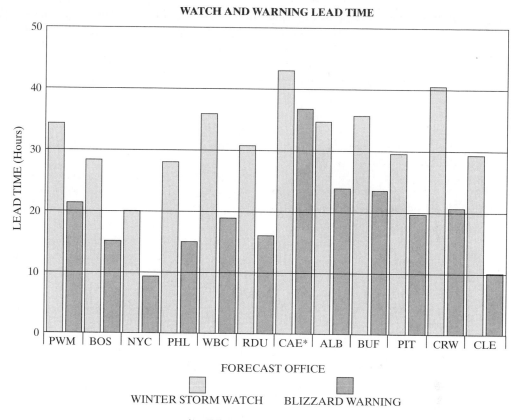

Figure 6.7 Comparison of lead times for winter storm watches and blizzard warnings (heavy snow warnings were issued at Columbia, SC (CAE)) issued by the local NWS Weather Forecast Offices in the Eastern Region: RDU—Raleigh-Durham, NC; WBC—Sterling, VA; CRW—Charleston, WV; PHL—Philadelphia, PA; PIT—Pittsburgh, PA; CLE—Cleveland, OH; NYC—New York City, NY; BUF—Buffalo, NY; ALB—Albany, NY; BOS—Boston, MA; PWM—Portland, ME. Lead times are in hours.

week to 10 days into the future. In fact, theory suggests that the limit to such deterministic predictability is about two weeks. However, although there is little or no predictability of specific weather events beyond two weeks, there are methods that have increasing skill in predicting general trends or climatic anomalies several weeks, months, or even a year or two in advance. These methods predict the large-scale features of the atmospheric general circulation, for example, the mean positions of the long-wave troughs and ridges. For these longer time scales, anomalies in the temperature of the ocean play a role in forcing the large-scale atmospheric patterns.

For time periods ranging from a week to a month or two, global general circulation models which use observed sea surface temperatures show some small skill in predicting the mean position of the long waves in the atmosphere, and hence are useful in estimating where anomalous warm and cold and dry and wet conditions will prevail. For time periods beyond a month or so, it is necessary to consider changes in the ocean surface temperature, and thus *coupled* ocean-atmosphere models are used for long-range predictions. In these coupled models, the atmosphere and the oceans interact through wind stresses and exchange of heat and moisture at the interface between the atmosphere and the ocean. As discussed in Chapter 7, coupled models are showing increasing skill in predicting the *El Niño* phenomenon and associated global anomalous circulation patterns 6–18 months in advance. These forecasts can be of enormous use in long-range agricultural planning. For example, if a wet season is forecast farmers might plant rice, which thrives on wet conditions. Conversely, if a dry season is forecast, farmers might choose to plant cotton, which is tolerant of dry weather, instead.

6.3   WEATHER MODIFICATION

Humans have been modifying their atmospheric environment since they first lit fires and moved into caves. But even outside of their shelters, humans have been modifying weather and climate both intentionally and unintentionally, for a long time. They have done it by changing the contours of the land, changing the surface properties, and contaminating the air. But until recently, all of these changes have been done on a relatively small scale. As will be discussed in Chapter 8, human activities have now increased to the level where they are affecting climate on larger scales. Here we briefly review attempts to intentionally modify the weather.

The various scales of the weather patterns must be kept in mind when one considers the feasibility of any particular scheme for deliberate **weather modification**. On a large or medium scale, we cannot hope, at least in the foreseeable future, to change the climate by inputs of energy equaling those of natural atmospheric processes. We could hardly match the rate at which heat energy is converted to kinetic energy even in a small thunderstorm, and the total rate of energy conversion increases greatly as the circulation size increases. (See Problem 1, Chapter 5.)

On a small scale, there are a few examples of the direct use of heat energy to change the weather. Heating of the air over crops to save them from frost damage has been practiced for a long time. Generally, however, the produce must have a fairly high value, such as citrus fruit has, to make the practice economically feasible. During World War II, fog over English airports was sometimes dissipated sufficiently to allow aircraft operations by burning oil to raise the air temperature locally a few degrees above the dew point. In some places sidewalks are kept clear of snow by running hot-water or steam pipes beneath the pavement. In the densely populated areas of New York City during the winter, the thermal energy released to the atmosphere while heating buildings and generating power actually exceeds the amount received from the sun to ground level. But in these examples, we are dealing with releases of rather large amounts of heat energy over relatively small areas. These energies do affect the climatic conditions of these limited areas. But when considered over larger scales, human input of energy into the atmosphere is minuscule. For example, the total electrical energy produced each day in the United States is equivalent to just the latent-heat energy that is released during precipitation of 2.5 millimeters (0.1 inch) over an area 50 kilometers square (an area about half the size of Rhode Island).

Even though it seems unlikely that humans have used or soon will use "brute force" to modify their atmospheric environment, except on a very small scale, we might change the weather in either or both of two ways: (1) by altering the existing "natural" energy exchanges that occur among the Earth's surface, the atmosphere, and space, or (2) by stimulating or "triggering" various forms of instability that sometimes arise in atmospheric processes. These two "methods" are not mutually exclusive, of course, since a change in the energy balance may trigger instabilities, and the setting off of instabilities may release significant amounts of energy. An example of interference with the natural energy balance is the change that occurs when a city is built on what were green fields. An example of triggering atmospheric processes is **cloud seeding**: supercooled water droplets in a cloud may be induced to freeze, grow at the expense of the remaining water drops in the cloud, and precipitate; at the same time, the release of the latent heat of fusion may provide energy for additional vertical growth of the cloud.

Modification of clouds by the use of dry ice and silver iodide has already been mentioned in Chapter 1. Cloud seeding has not been proved to produce significant changes in precipitation amounts, except under certain favorable circumstances over relatively small areas. In fact, it has been much more effective in the **dissipation** of cold (less that 25°F) fogs than as a means for stimulating the rainfall rate. Figure 6.8 shows the effects of seeding a cloud from the air.

Most weather changes that have been produced or proposed are those that result from changes induced in the composition of the air or clouds, or in the surface properties of the Earth. One example of a change in surface properties that produces a change in the microclimate is that caused by the extensive areas of asphalt and concrete in a city. Water is often used to control the temperature over crops. On a clear, cold night, the temperature over an irrigated field will be noticeably higher than over dry soil.

Figure 6.8   Cloud deck seeded from the air. Note where cloud has been dissipated by seeding. (Courtesy of AFCRL.)

TABLE 6.2   Detailed station plotting guide for Figure 6.1

Station Number	Data Plot	Station Number	Data Plot	Station Number	Data Plot	Station Number	Data Plot	Station Number	Data Plot
1	22 / 13 — 209	11	12 / 6 — 256	21	24 * 32 — 179	31	13 / 2 — 256	41	61 / 58 — 837
2	43 / 18 — 200	12	27 / 15 — 284	22	−35 / −8 — 309	32	16 / 5 — 186	42	44 / 42 — 785
3	54 / 52 — 171	13	23 / 5 — 224	23	−8 / −14 — 304	33	30 ** 30 — 184	43	30 ** 28 — 977
4	58 / 43 — 156	14	−9 / −13 — 335	24	17 ** 9 — 211	34	45 — 143	44	38 ** 33 — 909
5	17 ** 10 — 279	15	14 / 8 — 312	25	27 / 5 — 199	35	27 ** 25 — 054	45	59 — 850 / 47
6	32 / 21 — 244	16	22 / 16 — 303	26	38 / 17 — 160	36	34 / 34 — 044	46	61 / 50 — 970
7	35 / 27 — 213	17	31 / 20 — 294	27	−25 / −33 — 273	37	30 ** 28 — 959	47	71 / 59 — 050
8	49 / 25 — 155	18	36 / 28 — 322	28	18 / 14 — 195	38	34 * ** 32 — 875	48	69 / 65 — 019
9	−19 / −*27 — 260	19	35 / 20 — 269	29	−12 / −23 — 199	39	64 / 59 — 943	49	62 — 040
10	−1 / −7 — 299	20	54 / 50 — 221	30	−11 / −17 — 264	40	71 — 961	50	33 / 24 — 241

Pollutants in the air affect the heat content of the atmosphere by reducing its transparency to the sun's incoming energy and the Earth's outgoing heat. Since dust particles serve as **cloud nuclei**, pollution leads to denser fogs and smogs. Fog is much more frequent over cities than over the surrounding countryside. It may also be that the greater number of nuclei leads to more precipitation over cities.

Control over the water supply can be achieved in ways other than by increasing precipitation. Suppression of evaporation from lakes and reservoirs is one technique that has been employed. This has been done by spreading a monomolecular film of a substance such as acetyl alcohol on the surface of the water. Evaporation can be retarded by 15–20 percent in this way, but it is difficult to prevent the film from being broken by waves.

Snow is a very important natural "reservoir" of water for many places. If the rate at which snow melts could be controlled, a steady supply of water might be available throughout the year instead of having an overbalance in the spring. Increasing the rate of snowmelt is not too difficult. The high reflectivity of snow can be decreased by covering the surface with some dark material such as lampblack. In Tibet it has long been the practice to throw pebbles on snow fields to speed melting for early planting. But no practical method for retarding snowmelt on a large scale has yet been suggested; slowing snowmelt would be immensely valuable to regions such as California, where most of the year's water supply comes in the form of snow over the mountains.

Various proposals have been made for weather modification on a grandiose scale. For example, it has been suggested that large areas of lowlands be flooded to temper the climate of surrounding areas. In northern Siberia the enormous annual temperature range (over 100°F) might be sharply reduced by extensive flooding.

A land bridge across the Bering Strait, cutting off the circulation of waters of the Arctic and Pacific Oceans, has existed in the past, and it has been proposed that this bridge could be rebuilt. This barrier would stop the flow of heat between these bodies of water, presumably raising the temperatures to the south of the strait and increasing the temperature contrasts across the barrier.

The question of weather control is a very important one and deserves serious consideration. But until the atmospheric scientist understands atmospheric processes more fully and therefore the possible effects that his tinkering may have, he or she must proceed cautiously. After all, we are very delicately tuned to our existing environment.

KEY TERMS

anticyclone	500-mb map	numerical weather prediction
chinook	fog dissipation	ridge
cloud nuclei	infrared satellite photograph	short-range forecasts
cloud seeding	isobaric surfaces	stationary front
cold front	jet streak	surface map
cyclone	jet stream	trough
deterministic forecasts	long-range forecasts	visible satellite photograph
extended-range forecasts	low-pressure system	weather modification

PROBLEMS

1. Identify the major ridges and troughs at 500 millibars in Figures 6.4–6.5. How do the sea-level cyclones and anticyclones move in relation to the large-scale wave patterns?

2. Determine the speed and direction of displacement of the cyclone on Figure 6.3 for the four 12-h periods (12 UTC Mar. 12 to 12 UTC Mar. 14).

3. Is the complexity of the frontal analyses on the sea-level charts greater over the oceans or over the continents? Why should there be a difference?

4. Examine the latest weather map published in your local newspaper. Predict whether there will be precipitation (and, if so, the type) and what the temperature, wind, and cloudiness will be in your city 24 hours from the time of the map. List the factors you took into account in predicting each element.

5. Some proverbs state that physical appearance of certain insects and animals is an indication of future weather. Do you doubt their validity? Why?

6. Compare the total energy received from the sun within the boundaries of your city on January 1 with that generated by power plants and heating units. (Use Fig. 3.8.)

7. Why is a "ring around the moon" a fairly good indication of an approaching storm? Can you think of a reason why the following weather rhyme might have a sound meteorological basis?

    Rainbow in the morning, sailor's warning

    Rainbow at night, sailor's delight.

8. During the middle of the afternoon on a quiet summer day, the temperature of the air is 82°F, the dew point is 57°F, and cumulus clouds are observed. How high are the bases of the clouds? What is the temperature at the cloud bases? What would be the height of the freezing level in the clouds (assuming that they extend that high)?

9. List all the ways in which you think humans may be affecting the climate on various scales (i.e., from areas as small as, say, your backyard, to regions as large as an entire continent), both intentionally and unintentionally. Can you find any evidence from weather records that the microclimate in your area has been changed by humans?

10. How did the temperature, winds, and heights change at 500 mb over Cleveland, Ohio between 7 p.m. E.S.T. Mar. 12 and 7 a.m. E.S.T. Mar. 14, 1993? [Compare Figures 6.4 and 6.5.]

11. Estimate the geostrophic wind speed (see equation 4.5) at a point in central South Carolina at 7 p.m. EST 13 Mar. 1993 using the weather map shown in Fig. 6.5a. Do the same for Chicago, Il. Why are these geostrophic wind speeds higher than the observed wind speeds?

7
Climate

7.1 THE NATURE OF CLIMATE

A common misconception is to think of the **climate** of a region as only the average state of the atmosphere. Not only do the average temperature, precipitation, wind, and other elements determine the climate, but their variations do also. The diurnal, day-to-day, and seasonal changes, as well as the extremes of the weather, are important in determining what crops can be grown, how homes and other buildings must be designed, and the way in which many other human activities may be conducted.

The major factors that control climate are: (1) the intensity of solar radiation, which is a function of latitude and season, (2) the reflectivity (albedo) of the Earth's surface, (3) the distribution of land and sea, (4) the topography. Many local influences affect the small-scale or **microclimate**—vegetation characteristics, small bodies of water such as lakes, and human activities that alter the surface properties or the composition of the air.

7.2 TEMPERATURE

The average as well as the diurnal and annual variations of temperature are determined principally by latitude, altitude, and the influence of land and sea. Some of these effects can be seen from the average world isotherms of Figure 3.10. The *latitudinal* variations can perhaps be seen more clearly from Table 7.1.

The mean temperature decreases and the range increases with latitude. But the difference in the percentage of land mass in the two hemispheres also shows up clearly. The average annual range for the Southern Hemisphere, which not only has less land mass but also has it concentrated in the tropics, is half that of the Northern Hemisphere.

TABLE 7.1   Mean annual temperature and temperature range and their variation
with latitude.

Latitude	Mean Temperature (°F)		Mean Annual Range (°F)	
	N. Hem.	S. Hem.	N. Hem.	S. Hem.
90–80°	–8	–5	63	54
80–70°	13	10	60	57
70–60°	30	27	62	30
60–50°	41	42	49	14
50–40°	57	53	39	11
40–30°	68	65	29	12
30–20°	78	73	16	12
20–10°	80	78	7	6
10–0°	79	79	2	3

The difference between the temperature regimes of stations close to the ocean and those well in the interior of continents is illustrated by Figure 7.1. Note how annual range increases with distance from the ocean shore, especially from the western shore, because the stations represented are generally within the belt of prevailing westerlies. (Hawaii is an exception.) The range of temperature is an index of what is called the **continentality** of a station. The **diurnal temperature variation** is also dependent on continentality, as can be seen from Figure 7.2.

The effect of altitude on the temperature range is illustrated by Figure 7.3. In parts of the elevated southwest, the average difference between day and night temperatures in winter is 33°F, while a few hundred miles to the east the range is about a third less. This effect largely reflects differences in moisture and cloudiness in the two areas.

Temperature Inversions.   Although the temperature typically decreases with elevation in the troposphere, there are important exceptions to the general rule. For example, temperature **inversions** along the west coast of South and North America at middle and low latitudes are persistent normal "climatological" features of the temperature distribution. Along the coastal hills the average temperature is often warmer at elevations of 1000 meters or so than it is near sea level. Drainage of cold air into low spots in mountainous terrain produces pockets of cold air throughout much of the year. Farmers know this, and they plant their frost-sensitive trees and crops along the slopes, leaving the bottom land for hardier plants.

Temperature Indices.   Human comfort is closely related to the body's heat budget. When the body loses heat faster than it is produced, or if the body loses heat more slowly than it is produced, it suffers discomfort and, under extreme conditions, injury or death. There are several factors that determine the body's rate of heat loss, but one is the temperature of its environment. Some idea of the amount of heating or cooling required in the buildings where people live can be obtained by determining the difference between the mean temperature of a day and some arbitrarily defined ideal temperature (usually 65°F). If one adds up all of these daily temperature

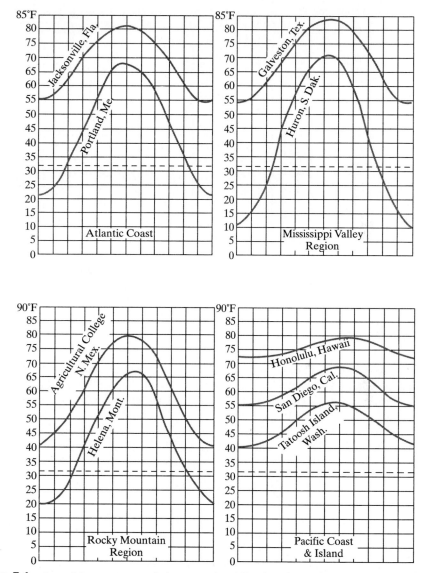

Figure 7.1 Annual temperature variation at continental and marine stations.

differences over, say, a month or a year, the result can be expressed in terms of **degree days**, a measure of how much heating or cooling will be needed to achieve ideal conditions. Heating engineers find such information useful in estimating fuel and equipment requirements for any locality. You can obtain a rough estimate of the temperature factors that affect your fuel bill from the average annual number of heating degree days in Figure 7.4.

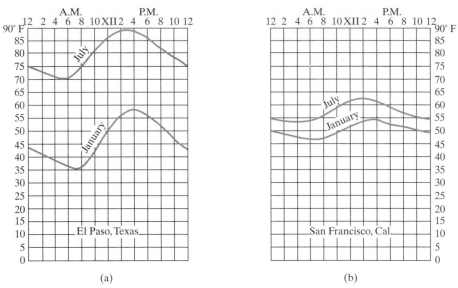

Figure 7.2   Diurnal temperature variations at (a) a continental, and (b) a maritime station.

7.3   PRECIPITATION

Average cloudiness and precipitation are linked most strongly to the general circulation and topography, and proximity to oceans and large lakes. Figure 7.5 illustrates the latitudinal variation of precipitation, which conforms, approximately, with the gen-

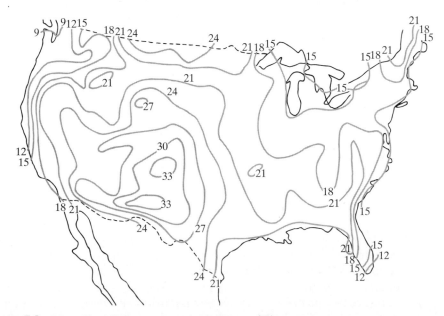

Figure 7.3   Mean diurnal temperature range in January (°F).

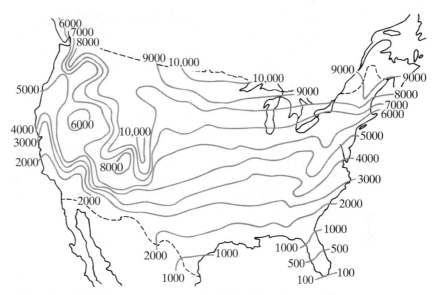

Figure 7.4 Annual heating degree days over the United States (base –65°).

eral circulation pattern of Figure 5.2. There is a peak in the doldrums belt where the trade winds converge. The amount drops in the subtropical anticyclone belt, but not drastically. Actually, it is along the eastern edges of the anticyclones in this belt that the great tropical and subtropical deserts are found; the western edges experience upward air motion and ample precipitation. Wave cyclones along the polar front produce much of the rain in the middle latitudes and high altitudes. The polar regions, because of their low temperatures, are quite arid.

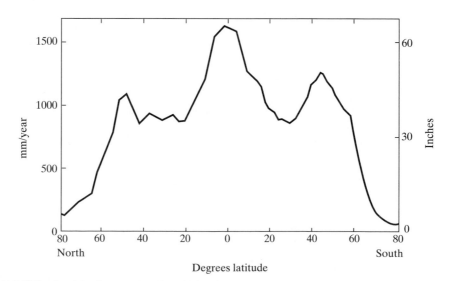

Figure 7.5 Precipitation as a function of latitude.

More rain falls on the oceans than on the land (Figure 2.7 and 7.6), and in the belt of prevailing westerly winds the west coasts of the continents have higher precipitation than the east coasts. Over islands the precipitation is greater than over the surrounding ocean, due to the orographic and convective (heating) effects.

Orographic precipitation (that induced by the forced ascent of air on the windward side of mountain barriers) is also an important factor in the rainfall distribution. Notable examples of orographically produced areas of high precipitation are found on the west side of the Rocky Mountains, on the west side of the Andes in central and southern Chile, and along the west coast of Norway. The rains of the summer monsoon over India and along the southern slopes of the Himalayas are greatly intensified by upslope motion. On the less side of mountain barriers, there are dry areas, called **rain shadows**. Examples of rain shadows are those found to the east of the Cascades in the states of Washington and Oregon and the arid Patagonia area in Argentina.

The seasonal distribution of rainfall has great significance, especially for agriculture. Precipitation is much more useful when it occurs during the growing season of plants than when it occurs at other times of the year. There are a great number of seasonal distributions of rainfall. Some of these are shown in Figure 7.7. Distribution (a) represents the equatorial type. There are two maxima that occur shortly after the equinoxes. These two maxima occur because the intertropical convergence zone oscillates over about 20 degrees of latitude during the year. Each time it passes a point near the equator, rainfall is increased. The other types are (b) tropical, (c) monsoon, (d) subtropical (west coast), (e) continental, and (f) maritime.

7.4   MICROCLIMATES

The climate over large areas frequently shows small-scale variations due to the effect of topographic features, vegetation characteristics, and even such human-made structures as buildings, roads, and artificial lakes or reservoirs. These microclimatic variations are sometimes very important in determining the crops that can be grown and may affect many aspects of our health and comfort.

The effect that Lake Michigan has on the microclimates in its vicinity is illustrated by Table 7.2. Grand Haven, Mich., which is generally on the leeward side of the lake, has a higher temperature, more precipitation, more days with snow, and less sunshine during the winter than does Milwaukee, which is on the windward side. During the summer, on the other hand, Grand Haven has a lower temperature and a slightly less precipitation than does Milwaukee.

Trees and plants produce microclimatic variations primarily because of their effects on the moisture supply and the wind flow near the surface. Transpiration from vegetation causes locally higher humidity, while the soil tends to inhibit precipitation runoff. Temperatures and wind speeds are lower within forested areas than in the open.

With continued proliferation of industrial and urban development, the importance of the microclimatic environment increases. The climate of large cities is characterized by increased "smog," dust, and waste gases from traffic, industrial sources, and domestic heating. While these pollutants prevent some of the sun's radiation

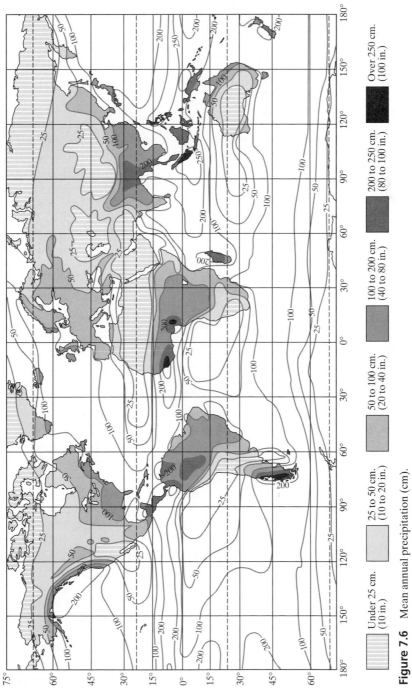

Figure 7.6 Mean annual precipitation (cm).

Under 25 cm. (10 in.)	25 to 50 cm. (10 to 20 in.)	50 to 100 cm. (20 to 40 in.)	100 to 200 cm. (40 to 80 in.)	200 to 250 cm. (80 to 100 in.)	Over 250 cm. (100 in.)

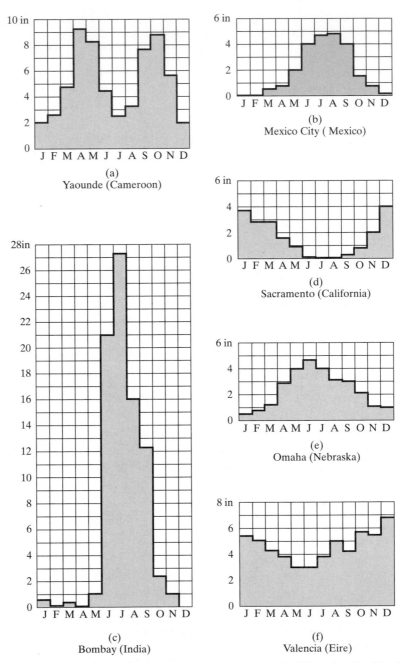

Figure 7.7 Annual distribution of rainfall: (a) Yacunde (Cameroon), (b) Mexico City (Mexico), (c) Bombay (India), (d) Sacremento (CA), (e) Omaha (NE), (f) Valencia (Erie)

from reaching the city itself, the decreased solar heating is more than compensated for by absorption and reradiation of the sun's energy by streets and walls of buildings, and by the lack of evaporative cooling, which would be present in forests and fields.

TABLE **7.2** Microclimatic difference between windward and leeward shores of Lake Michigan (after Landsberg).

Climatic element	Time of year	Milwaukee (windward shore)	Grand Haven (leeward shore)
Mean temperature	January		3.6°F higher
Mean temperature	August	2.0°F higher	
Precipitation	Dec.–Feb.		2.09 in. more
Precipitation	June–Aug.	0.55 in. more	
Snowfall	Jan.–Dec.		44 days more
Snowfall	Dec.–Feb.	20% more	

Furthermore, the buildings of the city tend to impede the natural flow of air. Consequently, the city center tends to be warmer than its surroundings and to suffer considerably more from impurities in the air. Recent studies also suggest that cities tend to have slightly more precipitation than the surrounding countryside.

By making use of his knowledge of microclimates, people can accomplish certain desirable changes in the environment. For example, removal of a thick cushion of moss was undertaken in the Yakut area of Siberia during World War II. Since the moss served as a heat insulator and also caused considerable evaporative heat loss, its removal permitted increased absorption of solar radiation by the ground. Consequently, the heretofore frozen soil thawed to a sufficient depth so that crops could be grown. Other examples of the use of microclimatic knowledge include the planting of rows of tall trees as windbreaks in the valleys of California and the designing of homes to take advantage of existing favorable microclimates produced by existing hills, streams, and vegetation in the local area.

Climatic Extremes.   Of considerable interest to almost everyone is "recordbreaking" weather, and newspaper headlines note the occurrence of a record low temperature or an excessively heavy snowfall at some location almost every year. Table 7.3 shows the highest and lowest values of a number of weather elements, representing extremes up to the current time. However, it is axiomatic that, by their very nature as extreme weather conditions, such records will continue to be broken.

7.5 CLIMATE CHANGE

Weather is noted for its variability. However, climate also changes. If one were to plot the value of some element of weather, such as temperature at a given location as a function of time, fluctuations over all time scales would appear. For example, if the air temperature were averaged over hourly intervals, these averages would change from hour to hour in response to the Earth's daily rotation. If these 24 hourly temperatures were then averaged for each day, one would notice day-to-day oscillations resulting from air-mass changes. Similarly, a distinct seasonal oscillation occurs because of the Earth's revolution around the sun. But fluctuations would also appear, due to inter-annual changes in the general circulation of the atmosphere. Even averages computed over decades and centuries fluctuate, as do averages over thousands and

TABLE **7.3**   Climatic extremes—highest and lowest observed values of certain meteorological elements.

Element	World Records	U.S. Records
Highest temperature	136° at Azizia, Libya, Sept. 13, 1922	134°F in Death Valley, Calif., July 10, 1913
Lowest temperature	–128.6°F at Vostok, Antarctica, July 21, 1983	–80°F at Prospect Creek, Alaska, Jan. 23, 1971
Greatest annual rainfall	1041.78 in at Cherrapunji, India, Aug. 1860–July 1861	739 in. at Kukai, Maui, Hawaii, Dec. 1981–Nov. 1982
Greatest monthly rainfall	366.14 in at Cherrapunji, India, July 1861	107.00 in at Puukukui, Maui, Hawaii, Mar. 1942
		71.54 in. at Helen Mine, Calif., Jan. 1909
Greatest 24-hour rainfall	73.62 in at Cilaos, La Reunion, Indian Ocean, Mar. 15–16, 1952	43 in at Alvin, Tex., July 25–26, 1979
Longest period without rainfall	19 years at Wadi Haifa Sudan (entire period of record)	767 days at Bagdad, Calif., Oct. 3, 1912–Nov. 8, 1914
Greatest 24-hour snowfall	Same as U.S.A. record	76 in at Silver Lake, Colo., Apr. 14–15, 1921
Greatest seasonal snowfall		1122 in at Paradise Ranger Station, Wash., 1971–1972
Highest wind speed at surface	Same as U.S.A. record	231 mi/h at Mt. Washington, N.H., Apr. 12, 1934
Largest recorded hailstone		17.5 in, 1.67 pounds at Coffeyville, Kans., Sept. 3, 1070
Greatest temperature changes in 12 hours		+83°F, Granville, N.D., Feb. 21, 1918 (–33°F to +50°F)
		–84°F, Fairfield, Mont., Dec. 24, 1924 (+63°F to –21°F)
Greatest temperature changes in 15 minutes		+42°F, Fort Assiniboine, Mont. from –5°F to +37°F
		–63°F, Rapid City, S.D., Jan. 10, 1911 from 55°F to –8°F

millions of years, judging from the indirect evidence available (Figure 7.8). Thus, weather is not a constant on any scale of time, but rather it appears to vary over all possible periods.

These cycles or periods (intervals of time over which the weather elements repeat themselves) are not generally very regular. Even the most regular of them—the daily and annual variations—are not unchanging. For example, the start and end of the summer monsoon over Southeast Asia vary considerably from year to year. In fact, forecasting from assumed periodicities in the weather has never been successful precisely because of the great irregularity of such cycles, even for periods much less than a year.

Systematic weather observations do not exist for longer than about 300 years, so long-period changes of climate must be inferred from evidence other than direct measurement. Figure 7.9 shows the contribution to the climate record from various types of direct and indirect data. Most of the indirect evidence gives information only

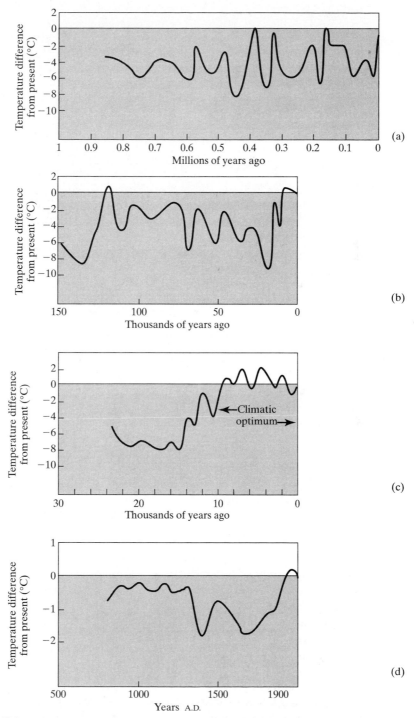

Figure 7.8 Variation in temperature (in °C) over the past (a) 1 million years, (b) 150,000 years, (c) 25,000 years, (d) 1200 years.

165

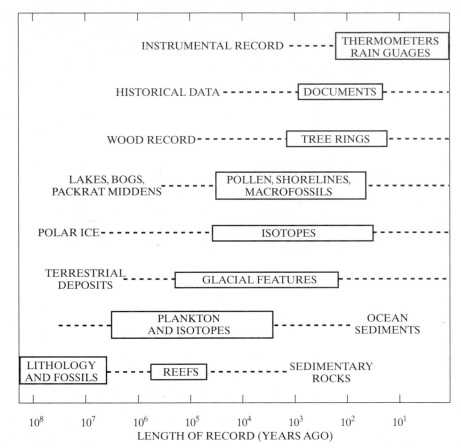

Figure 7.9   The time span of various climatic records from decades (thermometers) to hundreds of millions of years (fossils). (Source: Bernabo, J.C., 1978: Proxy data: Nature records of past climates. Environmental Data Service, NOAA.)

on precipitation and temperature. First of all, there are written records, which permit extrapolation backward in time for a few thousand years. Then there are the varying widths of growth rings of old trees, the migration of people, the fluctuations in levels of lakes and rivers, and the succession of plant types, which provide clues to the changes that have occurred during the past 10,000—20,000 years.

Geological evidence must be used to extend the time scale farther into the past. For this purpose, the types of fossil flora and fauna found in an area are indicative of the climatic characteristics at the time they lived. The advance and retreat of glaciers (snow fields), which leave their imprint on the Earth that they have traversed, are signs of changing climate.

Very little is known about conditions before 600 million years ago (before the Cambrian period), but great **ice ages** have occurred at intervals of about 250 million years: one some 700 to 1000 million years ago, another about 550 million years ago, a third 250 million years ago, and the latest, and best documented, which began roughly a million years ago and has only recently (geologically speaking) ended.

Figure 7.8 shows the variation of the Earth's mean temperature over the last million years. It is evident from these temperature estimates that only a tiny fraction of the past million years has had global mean temperatures as high as they are today. During most of this period the temperatures have been several degrees Celsius lower than at present, and huge climate changes have been associated with these apparently small changes in the mean temperature.

The oscillations in global mean temperature shown in Figure 7.8 have been accompanied by major advances and retreat of the ice sheets which have had enormous effects on the plant and animal life of the planet.

For example, in the last 700,000 years, there have been seven major advances and retreats of the ice. In the last of these advances, which lasted about 100,000 years, great ice packs extended hundreds of miles south of the Canadian border into the Mississippi Valley. Emergence from this last ice age began less than 20,000 years ago, with its completion only some 8000 years ago. At the peak of this glaciation, reindeer and musk-oxen penetrated the central United States, and walrus were found along the Georgia coast.

Beginning about 10,000 years ago, the European climate increased in warmth and dryness compared to the coldness and wetness of the preceding periods; between about 6000 and 3000 B.C., the climate reached an optimum—warm and humid—with annual mean temperatures about 2°C higher than today and with year-round rain. The European climate cooled somewhat and was rather dry between 3000 B.C. and 500 B.C. It cooled markedly between 500 B.C. and A.D. 200; during this period, the winters were snowy and cold and the summers cool and wet.

In this century, there is evidence that the global average temperature has shown a net increase of about 0.5°C or so, with a warming trend occurring from 1900 to 1942, a cooling trend from about 1942 to 1970 and a warming beginning in the 1970s and continuing through the 1990s (Figure 8.10). However, lack of reliable data over the oceans, particularly in the Southern Hemisphere, makes these trends somewhat unreliable. Small differences in the temperature estimates over the vast data-sparse regions of the Earth could alter these estimates of global temperature trends by a considerable amount.

The evidence for climatic fluctuations over all scales of time is conclusive. It has been shown in earlier chapters of this book that there are numerous factors that play roles in determining the state of the atmosphere: the energy received from the sun and its latitudinal distribution, the transparency of the atmosphere to solar and terrestrial radiation, mountain barriers, the frictional drag of the Earth's surface. Just as there exist a variety of causes for daily, day-to-day, and week-to-week changes of weather, so too it appears that there must exist changes in one or more of these factors that were responsible for the longer period, although irregular, changes of the Earth's climates. The possible causes for these long-period climatic variations can be grouped into three broad categories: (1) **astronomical** (anything that alters the amount, type, or distribution of solar energy intercepted by the Earth), (2) **atmospheric composition** (anything that changes the transmission or absorption of radiation by the atmosphere), and (3) **Earth's surface** (anything that alters the energy flow at the Earth's surface or changes its geographic distribution). Some causes for long-period climatic changes within each group are listed in Table 7.4.

TABLE 7.4   Possible causes for climatic change.

Cause	Approximate Range of Periods Induced (years)
1. *Astronomical changes:*	
A. Solar aging	10^9
B. Passage of solar system through galactic dust	10^8–10^9
C. Solar-output variability	10^1–10^8(?)
D. Earth-orbit changes	10^4–10^5
2. *Atmospheric composition changes:*	
A. Volcanic dust in the stratosphere	10^0–10^8(?)
B. Carbon-dioxide-content changes due to natural causes	10^4–10^8
C. Carbon-dioxide-content changes due to recent industrialization	10^1–10^2
D. Changes of other gaseous constituents	10^1–10^9
E. Dust particles introduced by human activities	10^0–10^2
3. *Earth-surface changes:*	
A. Migration of poles	10^7–10^9
B. Continental drift	10^7–10^9
C. Lifting of mountains and continents	10^7–10^8
D. Changes in relative sizes of ice caps and oceans	10^4–10^5(?)
E. Slow ocean circulation from great depths	10^3–10^6(?)
F. Slow adjustments between atmosphere and ocean	10^0–10^3
G. Changes in vegetation, e.g., deforestation	10^0–10^9

It should be emphasized that no single factor is responsible for *all* of the observed changes in climate. All causes have had *some* effect; the problem is to determine which are most significant. Specific causes have been postulated to cover almost every observed period of fluctuation. We shall comment on only a few of the theories.

Astronomical

The energy output of the sun, both the quantity and spectral distribution, varies with time. Recent measurements indicate that normal variations in the *total* energy output are about ±0.1 percent. However, there are intermittent outbursts of corpuscles (charged particles) and very short wavelengths (ultraviolet) of radiation associated with disturbances on the sun that are largely absorbed in the outermost layers of the atmosphere. These outbursts are known to affect the ionosphere and the Earth's magnetic field and to produce auroras, but it is difficult to understand at the present time how these very short waves of radiation, absorbed by the very thin atmosphere above 30 or 40 kilometers, can appreciably affect the great mass (more than 99 percent) of the atmosphere that lies below, where the weather occurs.

These outbursts from the sun are prevalent during years of relatively high frequency of sunspots. Sunspots have been observed and counted since the days of Galileo. Well-defined oscillations in the number of spots appearing on the face of the sun occur over periods of about 11 and 23 years. Although periodicities of about the same lengths have been found for certain elements of weather, such as air circulation, pressure, and temperature, it is difficult to say whether such correspondence is merely accidental without knowing how the phenomena might by physically linked.

The details of how changes in solar output might affect the climate are uncertain. It would appear unlikely that the effect would be merely to increase or decrease the average world temperature. Rather, an increase in solar output would result in greater heating in the equatorial regions than in the polar regions, thus increasing the latitudinal temperature gradient and therefore intensifying the atmospheric circulation. Stronger winds and higher temperature would likely lead to increased evaporation, clouds, and precipitation. Glaciation and the polar ice caps could either increase or decrease, depending on whether the effect of increased winter snowfall was more or less than the effect of warmer summertime temperatures. Little is known about the variation in solar output during the Earth's lifetime of about 5 billion years, although theories of solar evolution suggest that the sun's total output when the Earth was formed was approximately 30 percent less than it is today. Until there is some direct evidence of short-term changes in the sun's energy output, it will be difficult to either accept or discount this theory as a significant cause of climatic change over relatively short (thousands of years) periods of time.

A second theory in this group of astronomical causes regards Earth-orbit changes. As discussed in section 3.4, there are three types of variations in the Earth's motions: (1) The **obliquity of the ecliptic** (i.e., the angle between the plane of the Earth's orbit and the equatorial plane), which is now about 23-$\frac{1}{2}$ degrees, has changed by about $2\frac{1}{2}$ degrees in 45,000 years. When the angle is large, the seasons are more extreme and the pole–equator temperature difference in winter is increased. (2) **Perihelion** (the point in the Earth's orbit where the Earth is closest to the sun) now occurs during early January, but the date of its occurrence advances through the year with a period of about 21,000 years. (Thus, in 10,000 years, perihelion will occur in July.) When winter in one hemisphere coincides with perihelion, it will be a little milder than when it coincides with **aphelion**. This variation is called the **precession of the equinox**. (3) The eccentricity (**ellipticity**) of the Earth's orbit changes over a period of about 85,000 years. Although the difference in radiation received from the sun at perihelion compared to that received at aphelion is now only 7 percent, with maximum eccentricity the difference is as much as 20 percent.

Although these orbital changes only slightly affect the *average* yearly radiation received by the Earth, by changing the seasonal distribution they significantly alter the summertime melting and shrinkage of ice caps. Indirect evidence from ocean deposits confirms that orbital changes are significant producers of climatic change; however, they can hardly stand alone, since they cannot explain either the relatively short-period variations (less than 20,000 years) that have been observed or the long period without an ice age that occurred just before the last million years.

Atmospheric Composition

Volcanic eruptions sometimes spew fine **dust** particles into the upper atmosphere, and the particles can persist for years. These dust particles deplete some of the sun's rays by increasing the amount that is scattered by the atmosphere. In other words, increased dust causes an increase of the Earth's reflectivity (albedo). Figure 7.10 shows a satellite photograph of the eruption of El Chichón volcano. The oval-shaped white cloud over Mexico is not a water cloud; it is composed of dust particles and sulfuric acid aerosols produced by the volcano. As this cloud spread out, a small increase in

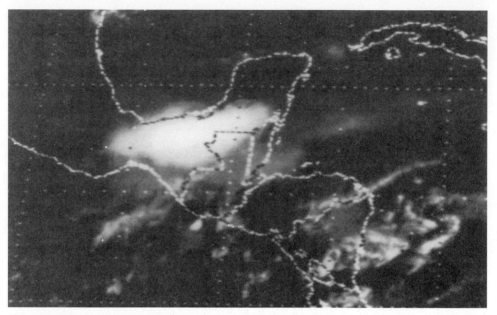

Figure 7.10   Satellite photograph shortly after eruption of El Chicón volcano in Mexico on March 28, 1982. A bright oval shaped cloud of dust is blowing downwind in upper tropospheric and lower stratospheric westerlies. (Source: NOAA Photograph.)

the Earth's albedo occurred over a large area. Even a small increase in albedo can be important; for example, a mere 1-percent increase in the Earth's albedo could lower the Earth's mean temperature about 1.5°C.

The eruptions of El Chichón in 1982 and Mt. Pinatubo (Figure 1.2) in 1991 produced large changes in the aerosol content of the stratosphere and upper troposphere. Figure 7.11 shows the atmospheric transmissivity in the visible wavelengths (0.3 to 2.8 μm) measured at Mauna Loa, Hawaii. (A transmissivity of 1.0 would occur with a completely clear (transparent) atmosphere; as the number of aerosols and hence the opaqueness of the atmosphere increases, the transmissivity decreases.) The rapid decreases in transmissivity in 1982 and 1991 clearly show the effects of El Chichón and Mt. Pinatubo. The effect of both volcanoes was to produce a global cooling effect of a few tenths of a degree Celsius, the effects of Mt. Pinatubo persisted longer than those of El Chichón, indicating a larger number of aerosols injected into the global atmosphere by Mt. Pinatubo.

As discussed in Chapter 1, trace gases such as water vapor, carbon dioxide, methane, and others are significant absorbers of long wave radiation emitted from the Earth's surface, and by reradiating some of this energy back to Earth, keep the surface of the Earth much warmer than it would be in the absence of these "greenhouse" gases. It follows that changes in the composition of the atmosphere over the history of the Earth has probably had profound effects on the climate. As will be discussed in Chapter 8, human activities are now causing relatively rapid increases in the greenhouse gases, and many scientists believe that these increases will cause significant global warming and other climate changes in the decades ahead.

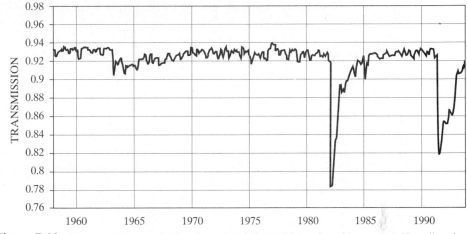

Figure 7.11   "Apparent" atmospheric solar transmission at Mauna Loa Observatory, Hawaii as determined from direct solar radiation measurements. (Source: NOAA.)

However, the dust or aerosol content of the atmosphere has also increased as a result of human activities, and the cooling effect of these human-produced aerosols tends to offset to a significant degree the warming effect associated with increasing greenhouse gases.

Earth's Surface

Anything that might alter either the properties of the Earth's surface or their distribution over the globe could lead to climatic change. Evidence from magnetic data in rocks (**paleomagnetism**) has shown that there may exist a slow wandering of the positions of the poles on the Earth's surface; i.e., in the past, the poles may have been at geographic points other than their present locations. The difficulty with this fact as an explanation of climatic change is that different areas would be affected at different times, but present evidence is that glaciers have occurred almost simultaneously over many regions of the Earth.

Continental drift across the globe, possibly induced by convection currents within the Earth's mantle, would certainly produce large changes in climate, but here again, they would not be simultaneous over many different land areas.

Mountain building is generally considered to be an important cause of climatic changes; especially those taking several-hundred thousand years or more. When a mountain is created, the increased elevation itself gives rise to lower temperatures; in addition the mountain reduces the exchange between polar and equatorial regions. (For example, the Himalayas are quite effective in preventing the mixture of warm and cold air masses to their south and north.) Mountain building and erosion take a long time and cannot explain the shorter periods of climatic change.

Expansion and contraction of the great polar ice caps would certainly have an important effect on world climates, principally because ice is such a good reflector (poor absorber) of solar radiation. Thus, an increase in the areas of the ice caps would decrease the total amount of solar energy by the Earth. It has been postulated, for example, that the great ice pack over Antarctica may periodically (about every 70,000 years?) slip into the adjoining oceans, spreading out to form continent-sized shelves. This slippage may occur when the ice depth over the Antarctic continent reaches some critical value such that the ice at the bottom which is under enormous pressure, begins to melt. (Heat flowing from the Earth's interior may help to melt this bottom ice.) If the ice were to extend itself over millions of square kilometers of ocean, the mean world temperature might decrease several degrees Celsius, and sea level would rise some tens of meters.

The oceans represent a tremendous reservoir of heat. Because the oceans cover such a large proportion of the Earth's surface, they are an important source of heat in the atmosphere; thus changes in the oceans' surface temperature could result in significant changes of temperature and moisture in the atmosphere. Not only can oceans affect climate directly through transfer of heat and moisture to the atmosphere, they can also produce indirect effects by altering the carbon dioxide content of the atmosphere. In addition to being soluble in water, carbon dioxide is taken up by microscopic plants and animals which live in the upper layers of the oceans. When these organisms die, their carbonate skeletons sink to the bottoms of the oceans. Thus life in the ocean acts to pump carbon from the atmosphere into the deep sea; the faster the pump operates, the less carbon dioxide remains in the atmosphere. Recent studies indicate that the ocean's control of atmospheric carbon dioxide, acting in response to changes in radiation due to orbital variations, is a major cause of climate change.

The ocean circulations operate over much longer time scales than do atmospheric circulations, hence the oceans have been called the "flywheel" of the climate. Figure 7.12 shows the global ocean circulation, which is called the **thermohaline circulation** because it is driven by difference in density associated with differences in ocean temperature and salinity. The thermohaline circulation acts as a giant conveyor belt transporting heat and salt for great distances as it modulates the global climate. In the Atlantic the thermohaline circulation is manifested as warm salty water transported by the Gulf Stream and North Atlantic current systems to high latitudes in the North Atlantic. There the salty water cools, becomes denser and sinks and return southward. If the atmospheric forcing at the surface were altered, for example by warming or freshening of the surface ocean layer from higher air temperatures and more precipitation, the water at high latitudes might become less dense and no longer sink, thus weakening the entire thermohaline circulation or causing it to break down completely. Such a change, which could occur on very short time scale (a few years), would have major impacts in the global climate.

Changes in the amount of type of **vegetation** covering the land can also have a significant potential impact on the climate because of the large effect of vegetation on the surface energy budget. Plants absorb more solar radiation than bare soil. Through evapotranspiration, they cool and moisten the air near the surface; it has been shown that irrigated portions of arid lands are much cooler and moister than nearby unirrigated and less vegetated lands. Hence changes in vegetation can certainly affect local and regional climates, and perhaps large-scale changes can affect the global climate. As discussed in Chapter 8, rapid deforestation in many parts of the world has raised concerns that significant and damaging changes in climate may result.

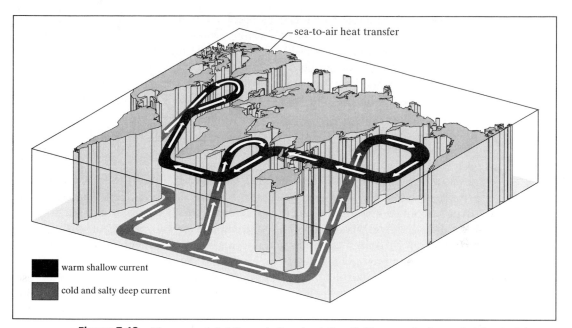

sea-to-air heat transfer

warm shallow current

cold and salty deep current

Figure 7.12   The ocean global thermohaline circulation. Colder water in the north Atlantic sinks to the deep ocean, to resurface and be rewarmed in the Indian and north Pacific Oceans. Surface currents carry the warmer stream back again through the Pacific and south Atlantic. This circuit takes almost 1000 years. (Source: Breecker, 1987.)

El Niño—An Example of the Natural Variability of Climate

Every few years, the normally cold, nutrient-rich water off the coast of Peru and Ecuador becomes warmer than normal. This warming was named El Niño by the fishermen of the region because it often appears around Christmas and Twelfth Night. Originally, El Niño was considered a very benign phenomenon by the people of this region because it brought rainfall and life to the barren shores of Peru; for this reason an El Niño year was known as *año de abundancia* (year of abundance).

With time, the perception of El Niño has changed for the worse, becoming associated with economic and ecological disasters. This has occurred because of economic development and huge increases in population. During an El Niño, the warm water kills the phytoplankton which is the basis for the food chain in the sea. Today the economy of Peru depends so heavily on the abundance of fish in the Pacific that their disappearance in an El Niño year is an economic disaster. Furthermore, the heavy rainfall which once was welcome now causes flooding and destroys the homes, roads and bridges that the large population requires. Thus, the extreme El Niño of 1982–1983 was associated with widespread flooding, loss of life and catastrophic destruction of property in Ecuador and loss of a generation of anchoveta fish.

Although El Niño has a severe local impact along the west coast of South America, it is part of a global-scale variability in climate which occurs periodically, though not regularly enough to be predicted reliably. El Niño of 1982-1982 may be used as an example of the characteristic climatic-anomaly patterns that are associated with these events.

In 1982, abnormally high sea-surface temperatures (as much as 4°C above normal) persisted into the winter season (Figure 7.13) and were associated with major perturbations in the global-scale pressure, wind,and precipitation patterns. Notable changes occurred in the sea-level pressure patterns in the Southern Hemisphere and the low-level winds in the equatorial Pacific (the normal easterly trade winds were replaced with weak westerly winds).

The warmer sea-surface temperatures and the anomalous westerly low-level flow were associated with a major shift in tropical precipitation patterns, with heavier than normal precipitation along the equator, and with severe drought conditions in Indonesia and Australia. The increased precipitation was responsible for a general warming of the troposphere by about 1°C. This warming implies an enhanced mean upper-level pressure gradient and hence faster than normal westerlies in the latitude band 15°–35°N. Over North America, this anomalous upper-air pattern resulted in a stronger than normal jet over the southern United States, Mexico, and the Gulf of Mexico during the winter of 1982–1983. Frequent, intense cyclones in association with this strong jet produced much heavier than normal winter precipitation over the southern third of the United States, creating unprecedented flooding in some areas.

Similar climate anomalies appeared over the Earth during the El Niño event of 1990–1993, which was one of the longest El Niños in history. Figure 7.14 depicts some of the significant weather and climate anomalies during 1993. While not all of them can be linked directly to this moderate but persistent El Niño, many of them are consistent with general circulation anomalies that occur during El Niño years. Notable events included devastating summer floods in Asia; much wetter than normal condi-

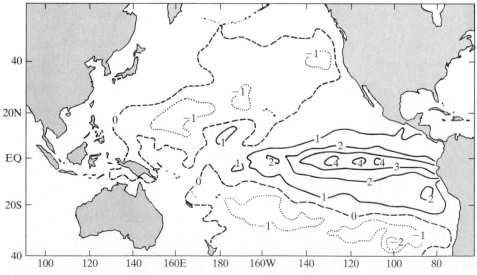

Figure 7.13   Sea-surface temperature anomalies (°C) for December 1982.

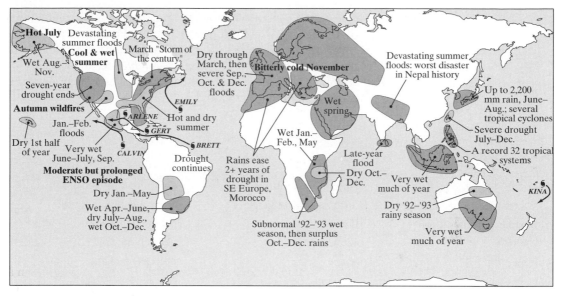

Figure 7.14 Significant climatic anomalies and events in 1993. (Source: Climate Analysis Center, NOAA.)

tions in Uruguay, southern Brazil and northern Argentina; record autumn cold in Europe; drought-breaking rains in California and one of the worst flooding events on record in the midwestern United States (see Figure 7.15).

Figure 7.15 Flood of 1993.

Considerable progress is being made in understanding the complex interactions between the ocean and atmosphere which produce El Niños. Predictive models of the coupled ocean-atmosphere system are showing skill in predicting El Niños and associated anomalous circulation patterns 6–18 months in advance, and there is increasing hope that these predictions will be useful to societies in many regions of the world in planning agricultural and other activities.

KEY TERMS

aphelion

astronomical climate changes

atmospheric composition
  changes

climate

continental drift

continentality

degree days

diurnal temperature variation

Earth's surface changes

El Niño

ellipticity

ice ages

inversion

microclimate

mountain building

obliquity of the ecliptic

orgraphic precipitation

paleomagnetism

perihelion

precession of the equinox

rain shadow

vegetation

PROBLEMS

1. Why is there so much snow and ice over the Arctic and Antarctic even though precipitation is light? Assuming the average precipitation over the Arctic is that of the average at latitude 80°N, what would be the minimum age of the ice at the bottom of a 30-meter iceberg? (Neglect compression of the snow.)

2. Why does the equatorial type of rainfall distribution have two maxima that occur shortly after the equinoxes?

3. Why does the maximum of precipitation occur in summer over the interiors of continents?

4. Explain the maxima of thunderstorms in Florida and over the Rockies (Figure 5.25).

5. From a knowledge of the general circulation, deduce the characteristics of the climate of the state of Washington, taking into account the topography.

6. Why are there relative minima of precipitation at latitudes 20°–30°S and N and maxima at 40°–50°S and N?

7. Suppose the obliquity of the ecliptic decreased from 23.5° to 10°. How would the seasons be affected? When would the winter solstice occur? How would the length of a June day in New York City be changed? A November day?

8. Suppose that much of the dry western United States were irrigated and the vegetation in this area increased. How would the climate in the summer be affected?

8

Global Change

July 1988 was the hottest month on record for the past 100 years in the Northern Hemisphere. Nearly 40 percent of the contiguous United States experienced severe or extreme **drought**. All counties in eleven states were officially declared disaster areas. The combined flow in the Mississippi, St. Lawrence, and Columbia rivers was 45 percent below normal. Only the dust-bowl years of 1934 and 1936 were drier. More than fifteen billion dollars in crop losses occurred and between 5,000 and 10,000 deaths in the United States were attributed to heat stress.

Extreme weather during 1988 was not confined to the United States, but rather was global in event and influence. **Heat waves** in Greece, India, and China caused 1,700 deaths; catastrophic **floods** in East and South Asia killed 10,000 people; heat waves and droughts in China, where rainfall was less than 25 percent of normal, caused 8,000 reservoirs and countless rivers to dry up; Hurricane Gilbert, the strongest Atlantic hurricane on record, caused massive property destruction in Haiti, Jamaica, and Mexico; while in Europe the mildest winter in 25 years saved 25 million barrels of oil.

Throughout the extreme heat of the summer of 1988, the notorious garbage vessel Pelican sailed around the world, seeking in vain a port that would accept its cargo of 14,000 tons of toxic incinerator ash that had been loaded onto the ship in Philadelphia in 1986. In its January 2, 1989 issue, *Time* magazine named the endangered Earth "Planet of the Year."

The Earth has assuredly always experienced extreme local weather and climate events; however, the magnitude and the global extent of the events of 1988 caused people around the world to take more notice of environmental issues than ever before. It became increasingly clear to many people that human beings, through their sheer numbers and extent of their activities, were changing the planet on a global scale at a rate faster than ever experienced by previous civilizations. Human activities are polluting the atmosphere and oceans, not only on local or regional scales as they have in the past, but now over the entire Earth. Through burning of fossil fuels, cutting of tropical and extratropical forests, disposal of toxic wastes, and construction of roads and

177

buildings, humans are changing the composition of the atmosphere, the characteristics of the land, the number and types of plants and animals, and the quality of rivers, lakes, and even oceans. Collectively, these changes are referred to as "**global change**." They threaten to alter the planet that has sustained such a variety of life for millions of years, in ways as great or greater than have ever occurred during its entire history.

Global change is not new. During the **Cretaceous period** (65–135 million years ago) changes in the environment were probably responsible for the extinction of the dinosaurs. Evolution and extinction of species are fundamental processes that have gone on for millennia. What is different about the present change from any previous ones is that it is caused by one species—humans—and the change is occurring at unprecedented rates. For example, the present rate of species extinction is 1,000 times faster than has prevailed since prehistoric times. This chapter discusses some of the causes and characteristics of global change, a phenomenon presently affecting all of life on Earth.

8.1 INCREASING HUMAN POPULATION: THE ROOT OF THE PROBLEM

It is axiomatic that the total impact, I, of the **human population** of the Earth's environment can be expressed as the total number of people living on the Earth (N) times the average impact caused per person, i:

$$I = N \times i \tag{8.1}$$

In the past 100 years, both N and i have been growing faster than ever before. Humans, who had only insignificant impact on the global environment during the one million years or so they had inhabited the planet, suddenly became major if not dominant players in its evolution.

To see how suddenly this has occurred, consider the growth of human population, depicted in Table 8.1. In the hunting and gathering period of human history, about 8000 B.C., there were approximately 5 million humans. By the time of Christ, this number had grown to about 200-300 million, and by 1650 A.D. to 500 million.

Not until 1850 A.D. did humanity reach one billion. It took almost 10,000 years to grow to one billion from 5 million in 8000 B.C. However, the second billion was reached in less than 100 years (1930), and the third and fourth billions in less than 50

TABLE **8.1** **Doubling times.**

Date	Estimated World Population	Time for Population to Double
8000 B.C.	Five million	1500 years
1650 A.D.	500 million	200 years
1850 A.D.	1000 million (1 billion)	80 years
1930 A.D.	2000 million (2 billion)	45 years
1975 A.D.	4000 million (4 billion)	35 years
1995 A.D.	5700 million (5.7 billion)	35 years

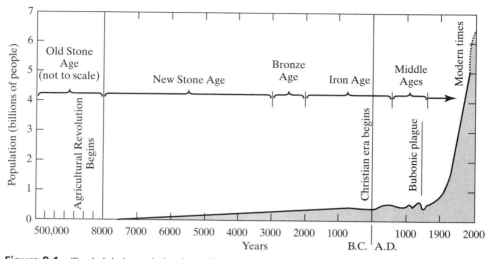

Figure 8.1   Total global population from 500,000 years before the present (BP) to the present (1990),
with an extrapolation beyond 1990. Note the change in time scale at 8000 years BP and at 1900. The drop
in population associated with the Bubonic Plague is greatly exaggerated on this figure; in reality even the
Plague resulted in only a small and temporary reduction in total population. (Source: Adapted from
Ehrlich and Ehrlich, 1972, *Population, Resources, Environment*, W.H. Freeman and Co., San Francisco.)

years. In fact, the doubling time for the total world population has shrunk from 1500
years to less than 35 years since 8000 B.C. When total global population is plotted on
a graph (Figure 8.1), the exponential nature of the growth is evident.

The growth of humans and their dominance in life on the planet is difficult to en-
vision. One measure of this dominance is especially graphic. Consider the total integrat-
ed mass of all vertebrate animals (mice, elephants, birds, snakes, etc.). Humans and their
livestock and pets are now 97 percent of the total mass, while all wild animals total only
3 percent. Ten thousand years ago, our human-related fraction was less than 0.1 percent.

Not only has the human population been growing faster than exponentially, the
average impact per individual has been growing as the average standard of living has
increased (though not exponentially). Thus, the total impact on the planet has been
compounded by both a rapid growth in N coupled with an increase in i in (8.1). One
measure of this growing impact has been the dramatic increase in the burning of fossil fuels over the past 100 years (Figure 8.2). Except for temporary slowdowns following World War I, the economic depression in the 1930s and the steep rise in oil
prices in the early 1970s, **fossil fuel consumption** has grown steadily worldwide. By the
late 1980s, fossil fuel in the United States was being used mostly for electrical power
generation by utilities (35 percent), followed by transportation (30 percent), industry (24 percent) and residential and commercial heating (11 percent).

Figure 8.3 shows the world's primary energy consumption by energy source and
by world region for 1985. Oil, coal, and natural gas are the primary sources of energy world wide, with biomass providing the major source of energy in developing countries. As shown in Figure 8.3, the industrialized countries, with 25 percent of the world's
population, use 66 percent of the world's energy.

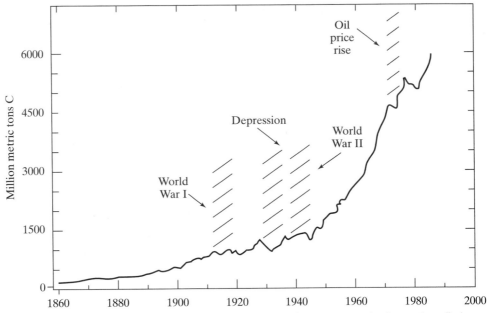

Figure 8.2 Total global CO_2 emissions from fossil fuel burning, cement production, and gas flaring, 1860-1988. (Source: *Trends* '90, U.S. Dept. of Energy.)

Another major impact of human population growth that has occurred mainly during the past 100 years is massive global **deforestation**. In Brazil's state of Rondonia, for example, 20 percent of the tropical rain forests have been destroyed since 1970. In 1988, between 20,000 and 50,000 square kilometers (an area equal to the size of Louisiana) of forests were burned in Brazil alone. Globally, the tropical rain forests are being destroyed at the rate of one football field per second, or 27 million acres per year! At this rate the tropical forests—home of between 50 percent and 70 percent of the planet's species—would be totally obliterated in 25 years. Apart from the impact on the regional and global environment, there would be an incalculable loss of biological and genetic diversity that would have fostered developments in medicine, agriculture, and biotechnology. A conservative estimate of the annual global loss of species due to deforestation is 5,000 species per year, more than 10,000 times the naturally occurring rate of extinction before humans existed.

Although the unprecedented rate of destruction of the tropical rain forests is receiving much attention from scientists and political leaders around the world, it is important to note that massive destruction of forests in temperature latitudes has already taken place and continues today. Great forests once covered North America from the Atlantic to the Mississippi. Giant sequoias in the Sierra Nevada, coastal redwoods in Northern California, Douglas firs in Washington and Oregon, and spruces and firs in British Columbia made up a nearly continuous forest in the Pacific Northwest. Now, on the continent as a whole, less than 5 percent of this virgin forest remains.

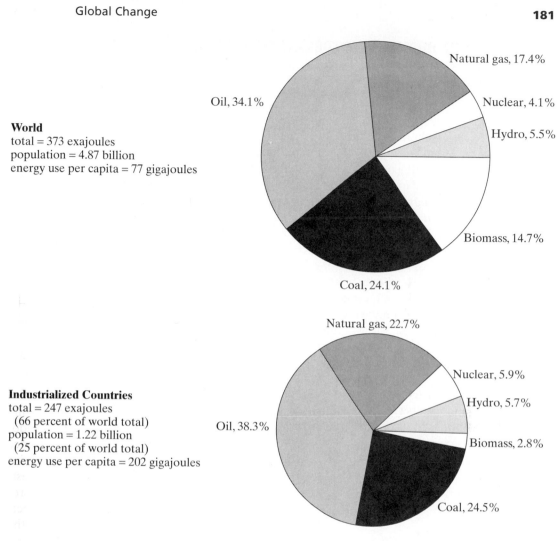

World
total = 373 exajoules
population = 4.87 billion
energy use per capita = 77 gigajoules

Oil, 34.1%

Natural gas, 17.4%

Nuclear, 4.1%

Hydro, 5.5%

Biomass, 14.7%

Coal, 24.1%

Industrialized Countries
total = 247 exajoules
  (66 percent of world total)
population = 1.22 billion
  (25 percent of world total)
energy use per capita = 202 gigajoules

Natural gas, 22.7%

Nuclear, 5.9%

Hydro, 5.7%

Oil, 38.3%

Biomass, 2.8%

Coal, 24.5%

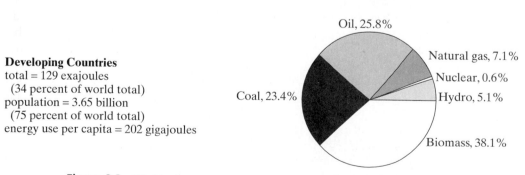

Developing Countries
total = 129 exajoules
  (34 percent of world total)
population = 3.65 billion
  (75 percent of world total)
energy use per capita = 202 gigajoules

Oil, 25.8%

Natural gas, 7.1%

Nuclear, 0.6%

Hydro, 5.1%

Coal, 23.4%

Biomass, 38.1%

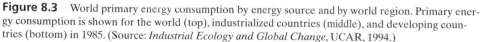

Figure 8.3   World primary energy consumption by energy source and by world region. Primary energy consumption is shown for the world (top), industrialized countries (middle), and developing countries (bottom) in 1985. (Source: *Industrial Ecology and Global Change*, UCAR, 1994.)

8.2 IMPACT OF HUMAN ACTIVITIES ON THE ATMOSPHERE

Increased Greenhouse Gases

The burning of fossil fuels and the associated release of **carbon dioxide** into the atmosphere have produced a steady increase in the average carbon dioxide content of the atmosphere. Figure 8.4 shows the concentration of CO_2 in parts per million by volume (ppmv) at a measuring station on top of Mauna Loa, Hawaii (in air that is far from local influences and is pristine). Superimposed on the annual cycle of CO_2 (minimum in the Northern Hemisphere summer when maximum plant growth sequesters CO_2 and maximum in the winter when plant growth is a minimum) is an unmistakable upward trend. Atmospheric CO_2 has increased by nearly 15 percent in the last 30 years and at a 1995 concentration of 358 ppmv is now higher than at any time in at least the last 160,000 years—the present extent of any available record. If half of our reserves of fossil fuel are consumed in the next 100 years, CO_2 concentration could increase above 1000 ppmv. As discussed in Chapter 1, CO_2 is an important **greenhouse gas**, affecting the radiation budget of the Earth. CO_2 and other greenhouse gases (mainly water vapor) keep the surface of the Earth 35°C (63°F) warmer than it would be without the greenhouse gases, other factors remaining the same.

Figure 8.5 shows the sources of carbon dioxide in the United States in 1977. Energy used for buildings, industry, and transportation contribute roughly a third each to the total CO_2 emissions.

Figure 8.6 shows the emissions of CO_2 by industry as a function of gross domestic product (GDP) per capita for 16 countries. The United States, as the largest national producer of CO_2 and with a high GDP per capita, stands out in the upper right corner of the figure with a per capita emissions of about 5.4 metric tons. Al-

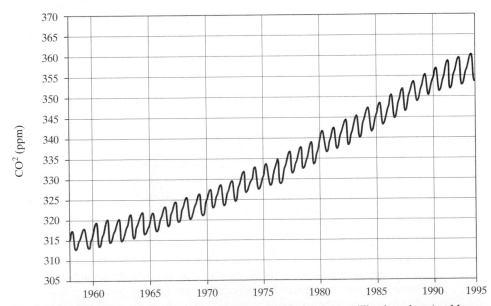

Figure 8.4 Monthly mean concentrations of carbon dioxide (parts per million by volume) at Mauna Loa, Hawaii. (Source: NOAA.)

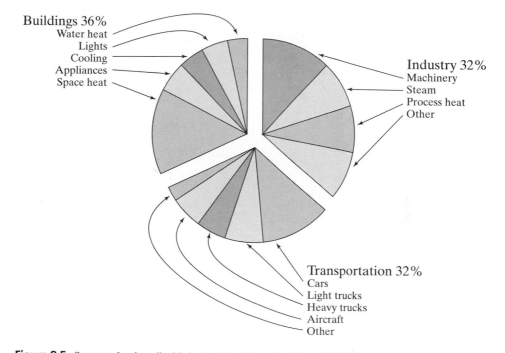

Buildings 36%
Water heat
Lights
Cooling
Appliances
Space heat

Industry 32%
Machinery
Steam
Process heat
Other

Transportation 32%
Cars
Light trucks
Heavy trucks
Aircraft
Other

Figure 8.5 Sources of carbon dioxide in the United States in 1987: total 1.3 billion metric tons. (Source: Office of Technology Assessment.)

though there is a correlation between CO_2 emissions and GDP per capita, the correlation is not perfect. "Efficient" countries like W. Germany, Japan and France have only slightly lower per capita GDP's than the United States, but CO_2 emission levels that range from one-third to just over one-half of the U.S. levels. In contrast, Poland and the former U.S.S.R. have per capita incomes comparable to those of Brazil, Mexico, and Venezuela, but CO_2 emissions several times their rates.

Although water vapor and carbon dioxide are the best known of the Earth's greenhouse gases, they are not the only important ones. Other trace gases, such as **methane, oxides of nitrogen**, and **chlorofluorocarbons** (CFCs) are also emitted by human activities such as industry and agriculture. Although these gases are less abundant than CO_2, some of them are much more effective greenhouse gases than CO_2. For example, methane, which is produced largely by agricultural activities (e.g. cattle raising, rice paddies) and by anaerobic decay, is present in much lower concentrations than CO_2 (1.7 ppmv compared to 358 ppmv), but molecule for molecule methane is 25 times more effective than CO_2 in trapping infrared radiation. A single molecule of a type of CFC (CFC-11) is 10,000 times more effective than a molecule of CO_2. These other greenhouse gases are also showing dramatic increases with time as a result of human activities. The United States led all other nations in producing greenhouse gas emissions in 1987 (Figure 8.7).

The rapid growth of methane in recent years is shown in Figure 8.8. Other greenhouse gases show similar trends. Taken together, it is estimated that the sum of all radiatively active trace gases except CO_2 presently produces an enhanced greenhouse effect that is about equal to that produced by CO_2 itself (Figure 8.9).

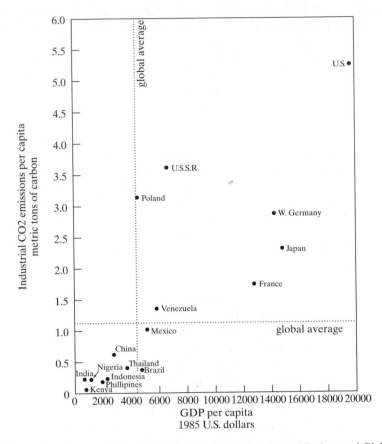

Figure 8.6 Static development plans for 16 countries. (Source: *Industrial Ecology and Global Change*, UCAR, 1994.)

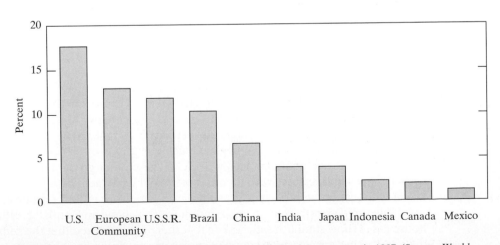

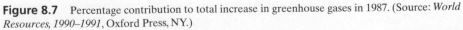

Figure 8.7 Percentage contribution to total increase in greenhouse gases in 1987. (Source: *World Resources, 1990–1991*, Oxford Press, NY.)

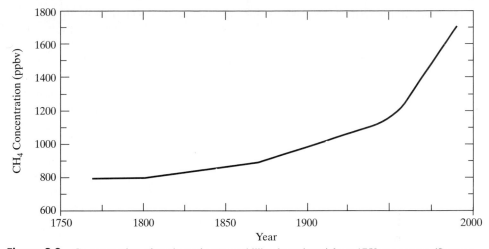

Figure 8.8 Concentration of methane (parts per billion by volume) from 1750 to present. (Source: Intergovernmental Panel on Climate Change. Climate Change: The IPCC Scientific Assessment, Cambridge University Press, 1990, NY.)

With the indisputable increase in greenhouse gases over the past 100 years, it would seem logical to suppose that there has been an equally indisputable increase in global mean surface temperature. This is not the case, however, for several reasons. First, the greenhouse effect is not the only important factor that determines the mean surface temperature and other aspects of climate, so that even if the greenhouse gases increase, other factors (such as cloudiness, dust in the atmosphere, changing **albedo** of the Earth's surface) could change in ways that for a time partially, or even completely, offset the enhanced greenhouse effect. Second, even if the climate-forcing functions (solar output, cloud cover, composition of the atmosphere, etc.) remained constant, the atmosphere can undergo natural oscillations on time scales

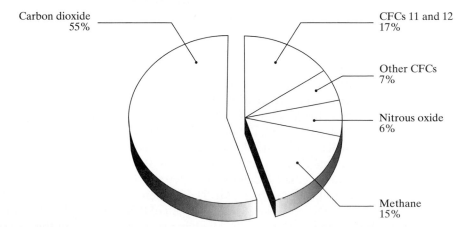

Figure 8.9 The contribution from each of the human-made greenhouse gases to the change in radiative forcing from 1980 to 1990. The contribution from ozone may also be significant, but cannot be qualified at present. (Source: IPCC.)

ranging from a few days to many decades, and global mean temperature fluctuations of 1°C or more can be associated with such oscillations. Finally, it is very difficult to measure the global mean surface temperature accurately. Temperature measurements, which often contain errors, are made from many sources over land and relatively few over oceans, and these must be combined to estimate the global average. Temperature measurements may be affected by very local heat sources (for example, in the middle of a city), and hence not represent the surrounding areas. And, of course, there are few or no observations over many regions of the globe, particularly vast regions of the oceans and in the Southern Hemisphere.

Taking the above factors into account, Figure 8.8 shows our best estimate of how the global mean surface temperature has varied over the past 100 years. A slow warming trend is evident from 1860 to about 1940. This was followed by a leveling off of the temperature change, or even a slight cooling, from about 1940 through the late 1970s. The decade of the 1980s brought a return to the gradual warming trend, and 1990 was the warmest year in the past century. The global cooling of 0.1 to 0.2°C in 1991–1993 is thought to be due to the eruption of Mt. Pinatubo in 1991. Overall the global mean temperature is estimated to have increased by about 0.5°C from 1860 to 1990. Although this amount of increase cannot be ascribed with certainty to the enhanced greenhouse effect for the reasons indicated above, it is consistent with theoretical estimates from models that incorporate the observed increase in greenhouse gases. It is also consistent with the marked recession of the majority of mountain glaciers over the same period. Glaciers worldwide have receded 11 percent over the past 150 years, and as much as 50 percent in some areas.

The observed rate of temperature increase shown in Figure 8.10 is also consistent with the estimated rise in global mean sea level over the past century (Figure 8.11). This rate of increase of between one and three millimeters per year is presumably caused by melting of the globe's continental **glaciers** and **polar ice** and thermal expansion of water.

The Ozone Hole

Ozone, a highly reactive form of oxygen that is present in minute amounts in the stratosphere, is essential to life on Earth because it absorbs biologically damaging ultraviolet light that is not so effectively absorbed by any other gas in the atmosphere. In fact, the evolution of the ozone layer between altitudes of 10 and 35 km was probably crucial to the development of life on Earth. Of immediate importance to humans, the ozone layer protects us from most of the solar radiation that causes skin cancers. In the United States, more than 500,000 skin cancer cases are diagnosed annually. Of these, a small fraction—approximately 27,000—are a deadly form called malignant melanoma, which kills more than 5,000 people in the United States every year. It is estimated that each 1 percent loss of total ozone would cause an increase in the total rate of **skin cancers** by 3 to 6 percent. In the United States this could mean an additional 40,000 cases per year. The incidence of **melanoma** increased by 83 percent between 1982 and 1989, although much of this increase is probably due to changing habits of dress and lifestyles involving more time spent in the sun rather than a decrease in global ozone.

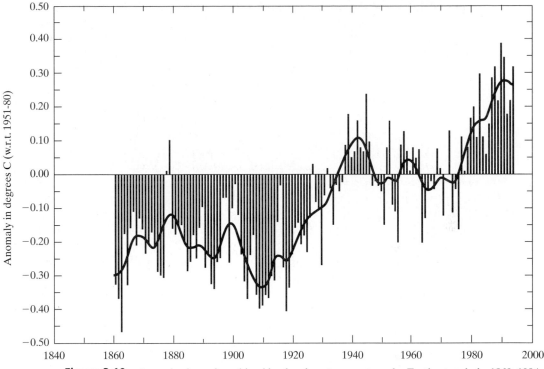

Figure 8.10 Annual values of combined land and sea temperatures for Earth as a whole, 1860–1994, shown as a departure in °C from the mean temperature for the period 1951–1980.

In the early 1970s, atmospheric chemists Sherwood Rowland and Mario Molina suggested that human-made CFCs could destroy ozone in the **stratosphere**. Inert in the troposphere, CFCs have lifetimes of 40–150 years. In the stratosphere, however, they undergo **photodissociation** and produce **chlorine** atoms, which can react with and destroy ozone molecules. Because of their chemical inertness and volatility, CFCs are popular for use as aerosol propellants, refrigerants, cleaning solvents for electronic components, and foaming agents for plastics. Ironically, CFCs were touted for years as being ideal chemicals for these purposes because of their inertness in the troposphere. However, this lack of chemical reactivity which makes them so useful also allows them to persist for many years. During this time they can make their way by vertical motions from the surface of the Earth into the stratosphere, where they can finally be dissociated by solar ultraviolet radiation, releasing ozone-destroying chlorine in the process.

Strong observational evidence that chlorine from CFCs could indeed destroy significant amounts of ozone came in 1985 when scientists from the British Antarctic Survey reported a 50-percent drop in total ozone present in the Antarctic stratosphere during the Antarctic spring season compared to the long-term average in this season. (Ironically, significant losses had been inferred from satellite observations prior to 1985, but the observations showed values that were so low they were reject-

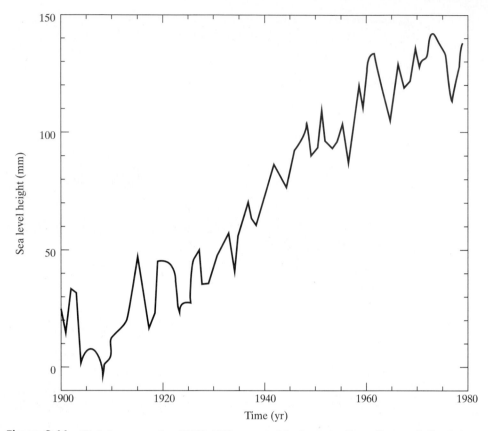

Figure 8.11   Global mean sea level 1900–1980, corrected for isostatic effects. (Source: A. Trupin.)

ed as measurement errors.) Figure 8.12 is a map of satellite measurements depicting this loss, quickly dubbed the "**ozone hole.**" Subsequent scientific expeditions to the Antarctic in the springs of 1986 and 1987 demonstrated unequivocally that the loss of ozone over the Antarctic was a consequence of chlorine chemistry. However, the process was somewhat more complicated then the mechanism originally proposed and was made possible by the very special meteorological conditions of the region. Over most of the world and during most seasons, the stratosphere is so dry that clouds do not exist, in spite of the cold temperatures of -50°C to -75°C. Over the Antarctic continent in winter, however, temperatures in the stratosphere at elevations of 15 to 20 km may fall as low as -90°C. At these low temperatures, some of the minute amounts of water vapor form ice clouds. Chemical reactions on the surfaces of the ice crystals composing these clouds convert chlorine from less-reactive forms to unstable forms that react in the presence of sunlight to destroy ozone. Because of the dependence of the process on sunlight, ozone destruction does not take place during the dark polar winter, but only after the return of the sun in spring.

The strong **circumpolar circulation** in the Antarctic stratosphere is another important factor in the ozone hole development. The intense, doughnut-shaped **polar**

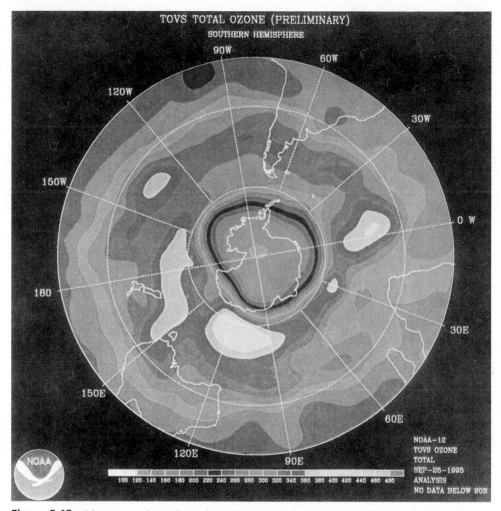

Figure 8.12   Map centered over South Pole of total ozone on October 12, 1993. Ozone hole (minimum in ozone) is centered over Antarctica, with values of ozone about 100 Dobson Units (100 Dobson Units corresponds to a layer of ozone 1 mm thick). Typical values of total ozone in this region before development of the ozone hole were 300 Dobson Units. (Source: NOAA.)

vortex isolates the air in the center of the vortex from warmer, ozone-rich air outside the vortex—in effect creating a natural chemical reactor. When the polar vortex breaks up in late spring, pools of ozone-poor air migrate away from the pole and can reach southern Australia, New Zealand, and the southern tips of South America and Africa.

Conditions over the **North Pole** are sufficiently different from those over the **South Pole** that the massive ozone destruction observed over Antarctica has not been observed over the North Pole, although some decreases (on the order of 10 percent) have occurred. The weaker and less symmetric north polar vortex, the higher stratospheric temperatures over the North Pole and the faster warmup in spring after sunrise appear to be the major reasons for the difference.

Observations of ozone in the nonpolar regions of the stratosphere have indicated a small (a few percent) loss during the wintertime in northern latitudes with little change during the summertime. Many scientists, however, regard the enormous losses found over **Antarctica** to be a global "early warning" of future changes that could be of great significance to life. In fact, many nations of the world have taken this warning seriously. In the so-called "**Montreal protocol**" of 1987, more than 30 nations agreed to a 20-percent reduction from 1986 to fully halogenated (the most harmful type) CFCs by 1994 and a further reduction of 30-percent by 1999. However, given the relatively small reductions in emissions and the long lifetime of CFCs already in the atmosphere, this protocol is likely to have only a minor effect over the next few decades in slowing the rate of ozone loss.

8.3   THE FUTURE

Because of the many factors that determine it, the natural (no significant effect from humankind) climate would be difficult enough to predict. When human influence becomes important, it exacerbates the problem of climate prediction, because assumptions about the growth of human population and human behavior must be made. In a system as complex as the climate, perhaps the best prediction is that there will be more surprises (no one predicted the ozone hole, for example).

Predictions of Global and Regional Climates

Even though specific climate predictions similar to two-day weather predictions are not possible, it is nevertheless useful to consider possible scenarios of future climate change and try to assess the most likely of these. A key first step in developing a likely scenario is to make assumptions about population growth and energy consumption. This in itself is difficult, since it is generally impossible to predict catastrophic events such as war, epidemics, or natural catastrophies that can radically change global energy use.

As shown earlier, the world population is growing exponentially. The Earth's population is doubling every 35–40 years. If growth continued at that rate, the population would exceed a billion billion people 1,000 years from now. This implies 1,700 persons per square yard of the Earth's surface, including all land and oceans! Thus it is absolutely certain that this rate of growth cannot persist indefinitely; either humans will control their own numbers, or other, catastrophic means such as global warfare and/or disease will control growth.

Nevertheless, it is entirely possible that the population will more than double in the next 50 years, reaching more than 10 billion by the year 2040, well within the lifetime of most readers of this text. In spite of efforts to use energy more efficiently, this population growth, most of it in the less-developed world, is almost certain to increase consumption of energy and thus increase emissions of greenhouse gases and other pollutants. Hence a very reasonable assumption is that the concentrations of all greenhouse gases will be twice the present values of 50–100 years from now.

Using both simple one-dimensional radiation models and complex three-dimensional **computer models** of climate (similar to the global models used for daily weather prediction), we can estimate how the climate would change if the total amount of greenhouse gases were to double. Different models have given estimates of mean global warming at the surface of between 1.5°C (2.7°F) and 5°C (9°F) for an atmosphere with twice the present amount of CO_2. The numbers may seem small, but even the low end of the estimate would produce a global climate warmer than ever before experienced by human civilization, occurring over an extremely short time compared to the rates of previous climate changes. For comparison, over the last two million years, glacial and interglacial cycles have occurred on time scales of 100,000 years with an accompanying variation in mean global surface temperature of about 5°C. Since the end of the last ice age, about 8000 B.C. mean surface temperatures have fluctuated by about 2°C over time scales of 100 years or longer. Figure 8.13 gives a range of possible increases of mean surface temperature based on assumptions of future emissions of greenhouse gases and the response of the climate to the resulting changes in concentrations of these gases.

The range of mean temperature changes predicted by the models indicates that there could be a significant rise in mean **sea level** caused by the melting of ice over Greenland and mountain glaciers and the thermal expansion of water. Estimates of likely **sea level rise** by 2030 range from approximately 10 to 30 centimeters.

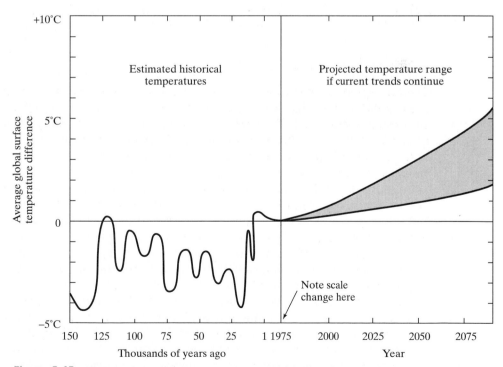

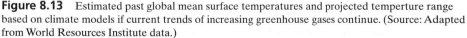

Figure 8.13 Estimated past global mean surface temperatures and projected temperture range based on climate models if current trends of increasing greenhouse gases continue. (Source: Adapted from World Resources Institute data.)

If the West Antarctic ice sheet were to melt completely, sea levels would rise by about 6 meters. A rise of this magnitude, not considered likely within the foreseeable future, would innundate virtually all of the world's major ports, including Boston, Miami, New York City, New Orleans, and San Francisco. Even a rise of 1 meter, considered by some likely to occur within the latter part of the twenty-first century, would have a significant effect on existing shorelines, particularly during extreme events such as storms.

It is very important to note that it is not the global mean temperature, or global mean value of any climatic variable, that is of most significance to life. Rather, it is the **regional climate** that humans and other ecosystems depend on and adapt to. Climate models and past data (paleodata) both indicate that regional and local climate changes are usually much more extreme than **global mean changes**; that is, when the Earth warms or cools on the average, changes in the general circulation, the jet stream, and storm paths cause large regional changes in precipitation and temperature. Some regions may even become cooler as the global mean temperature increases, and vice versa. Thus we expect that if the global mean temperature increases by several degrees, changes in certain regions will be considerably greater, perhaps as much as 10°C. Precipitation and evaporation are also likely to show large changes. Furthermore, the **frequency of extreme events** such as hurricanes, severe storms, droughts, floods, heat waves, and killing freezes is likely to change. In some regions, some of the changes are likely to be benign; for example, Canada and the Soviet Union might experience more productive growing seasons. These regional changes and frequency of extreme events are most likely to affect human society—indeed, all life. The regional patterns of temperature anomalies from one climate model shown in Figure 8.14 illustrates the large regional variability that these models predict.

Unfortunately, our best scientific tools for estimating regional climate changes and the frequency of extreme events—three-dimensional computer models—show much greater uncertainty than in their estimates of mean **global warming**. There are indications that the greatest warming will occur in high latitudes, that the midcontinents of middle latitudes will become drier, and that the frequency and intensity of hurricanes and typhoons will increase. However, because of the coarse spatial resolution of the models, and incomplete understanding of all the physical chemical, and biological processes and their feedbacks, these conclusions have a large degree of uncertainty.

It is also important to note that any warming associated with increased greenhouse gases is not likely to occur gradually and uniformly over the Earth. Changes in the other factors that determine the climate and the natural variability in the system will cause the climate to warm irregularly, making detection and response difficult. For example, even if the global mean temperature increases by 3°C in 50 years, a specific winter 50 years from now could be colder than a recent winter. The analogy of "loading the dice" is a good one; a greenhouse warming of several degrees will load the dice toward an increasing chance of warmer-than-present periods and a decreasing chance of colder-than-present periods. A mean temperature increase of 3°C in Washington, D.C., for example, would increase the average number of days with the temperature exceeding 100°F (38°C) from one to 12 per year.

Thus the hot summer of 1988 cannot be taken as proof of greenhouse warming, any more than an early snow or a cold winter of the 1990s (which is almost certain to happen) can be taken as proof that the greenhouse effect is not underway. Only time will tell for sure.

March, April, May

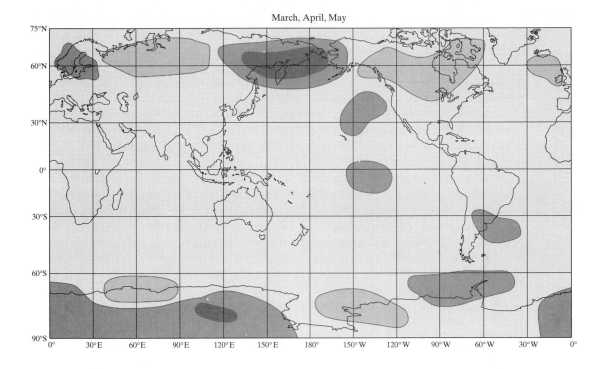

September, October, November

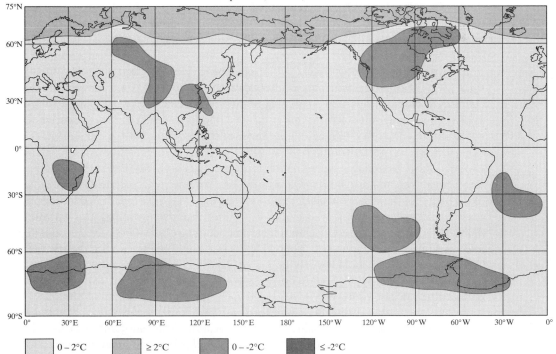

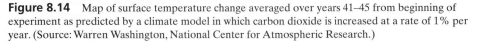

Figure 8.14 Map of surface temperature change averaged over years 41–45 from beginning of experiment as predicted by a climate model in which carbon dioxide is increased at a rate of 1% per year. (Source: Warren Washington, National Center for Atmospheric Research.)

However, on balance, the evidence is becoming greater that the human activities are causing a significant warming effect on the world's climate. The observed global warming over the past 100 years, the melting of the glaciers, the increase of water vapor in the atmosphere, the rising sea level, and computer model simulations are all consistent with the hypothesis that the increasing greenhouse gases associated with human activities are an important effect in the changing climate. In 1995, for the first time, an international panel of climate scientists working under the auspices of the Intergovernmental Panel on Climate Change (IPCC) concluded, "The best evidence to date suggests that global mean temperature changes over the last century are unlikely to be due to natural causes, and that a pattern of climate response to human activities is identifiable in observed climate records." The IPCC report goes on to suggest that global mean temperature will increase by an average of 2.5°C by the year 2100, which will trigger a 46 centimeter increase in sea level.

However, even if these estimates of future global changes are approximately correct, the changes will not be uniform over the Earth and will not occur gradually. Even with a mean global warming, there will be extremely cold winters and early and late snowstorms and freezes in many parts of the world.

Feedbacks in the Climate System

One of the difficulties in making predictions of climate change due to increased greenhouse gases, or another single factor, is the presence of positive and negative **feedbacks** in the climate system. When one variable (such as temperature) in the system changes, other variables (such as water vapor) may also change in response, and these secondary changes can either add to (positive feedback) or subtract from (negative feedback) the original change.

Although a consensus has developed among scientists that a significant (more than 1°C) mean global warming is likely during the next 50 years, and that regional changes in climate will be even greater, some scientists argue that too little is known about the complex interactions that produce climate, and that various negative feedbacks in the climate system will damp most of the greenhouse warming effect.

One of the greatest unknown feedbacks is that involving clouds and radiation. Clouds reflect incoming solar radiation, producing a cooling effect at the surface, but they also absorb and re-emit longwave radiation from the surface, causing a warming effect. The net of these two effects depends on the cloud height above the surface, latitude, season, and water content. Thus clouds can either produce a cooling or a warming of the Earth's surface. At present, clouds have a net cooling effect. Globally, the "**albedo effect**" produces a net cooling of about 45 watts per square meter (W/m^2), while the net "greenhouse effect" of clouds produces a net heating of about 30 W/m^2. The net cooling effect is not evenly distributed however, with most of it occurring over the mid- and high-latitude oceans (100 W/m^2) with the tropics being in a near state of balance.

The magnitude of this net **cloud forcing** (approximately 15 W/m^2) is about four times as large as the expected value of radiative heating from a doubling of CO_2. Hence small changes in the cloud cover over the Earth could play a significant role as a climate feedback mechanism. In fact, small changes in cloud cover could cancel completely the greenhouse warming effect. However, small changes could also enhance the greenhouse warming, perhaps doubling it. Our present knowledge about how global cloud properties would change in a warmer world is inadequate to choose between these two alternatives.

More certain is that the **water vapor-temperature feedback** loop is a positive one. Water vapor concentrations can be much greater in warm air than cold air because of the rapid increase of saturation vapor pressure with temperature. Water vapor is one of the most effective greenhouse gases, so as the lower atmosphere warms, water vapor will increase and will enhance the greenhouse effect. There is evidence that the global water vapor has increased by a few percent during the warm 1980s compared to the cooler 1970s, supporting the existence of this positive feedback mechanism.

Other important feedbacks include the **ice-albedo feedback**. If a warming Earth results in less ice over land and in the sea, the albedo of the Earth will decrease and more solar radiation will be absorbed at the surface (a positive feedback). This feedback is responsible for the greater warming seen at high latitudes in present climate models. However, if a somewhat warmer climate produces more snowfall at higher latitudes, snow and ice could increase, causing a negative feedback.

An important negative feedback that may well be occurring is the **CO_2 fertilization** effect. As CO_2 increases, the growth rates of many plants will increase because of enhanced photosynthesis and increased uptake of CO_2. There is evidence that this is occurring already; pine trees that in the past had a marginal existence are now growing well in California's mountains. Bristlecone pines have doubled their growth rates over the past 100 years. If more rapid growth occurs in response to increasing CO_2 in the atmosphere, and more CO_2 is in turn sequestered by the additional plants, a negative feedback slows the rate of increase of CO_2 and the greenhouse effect.

8.4   WHAT CAN BE DONE?

It should be evident from the discussion in this chapter that there is a potential for significant climate change, especially on regional scales, in the lifetime of our children. These changes could trigger serious international economic and political events. For example, worsened droughts in Africa and a continuation of the present rate of **desertification** there (more than 10 million acres of new desert are estimated to be formed each year, creating in only three decades new deserts the size of Saudi Arabia), coupled with that region's rapidly increasing population, could exacerbate existing social and economic instabilities. The spectre of massive numbers of environmental refugees seeking asylum in other parts of an already overcrowded and overstressed planet cannot be dismissed as impossible, or even unlikely. Given the changes many scientists and societal leaders believe possible, it seems only prudent to begin taking actions now to slow down the rate of environmental change. Even a 10 percent chance that some of the possible changes will occur warrants action, much as we as individuals take out insurance policies to guard against catastrophic accidents, illnesses, or economic misfortunes. Global-change skeptics may prove to be correct, but many believe that the risks are too great to do nothing while waiting to see what happens.

There are many actions that individuals and societies **could** take to reduce the possibilities of serious, even catastrophic problems that are conceivable due to a changing environment in the next century. Unlike other species, humans have the ability to understand global problems, to perceive threats to the global environment and to their very existence, and to conceive of the future. Many humans have already recognized these threats and have sounded alarms. In addition to their self-and world awareness, humans have the **technical capability** to effect solutions—to control their own numbers, to reduce their consumption of energy, and to recycle almost all of their waste. Thus, although political realities may make these solutions extremely difficult, some actions are at least **possible**.

Actions that could be taken that would reduce the risk of major environmental disasters over the next 50 to 100 years include:

Stabilization of World Population.   Almost all human-induced environmental problems are caused or exacerbated by increasing numbers of humans; almost all would be easier to cope with or solve if there were fewer rather than more humans. Thus the ultimate solution to environmental degradation of all kinds depends on a stabilization of the human population.

Conservation.   Reducing the per-person impact on the environment is an obvious mechanism to reduce the total rate of environmental damage. Developing more energy-efficient industries, transportation systems, appliances, and building heating and cooling systems represents an important potential component to the overall solution. Recycling of wastes conserves dwindling raw materials and alleviates the growing waste-disposal problem (Japan now recycles more than 50 percent of its waste.) Shifting lifestyles in affluent countries toward activities that use less energy (biking or walking rather than driving, for example) can make a contribution. Economic methods, such as energy taxes and requirements for more energy-efficient forms of transportation, may be necessary to achieve significant conservation.

Alternative Fuels, Renewable Energy Sources and Nuclear Power.   Alternatives to fossil-fuel sources of energy including solar power and safe nuclear power would conserve limited fossil fuels and reduce emissions of atmospheric pollutants, such as sulfur dioxide, in addition to greenhouse gases.

Efficient electric generation technologies could reduce emissions by about 600 million tons (30 percent). Nuclear power produces no emissions of greenhouse gases. In spite of the controversial nature of nuclear power in the United States, France generates more than 70 percent of its energy from nuclear power and has an impressive safety record. Renewable energy sources such as geothermal, photovoltaics, solar thermal systems, wind energy, and advanced biomass technologies are currently too expensive and produce too little energy to have a significant effect. Advances in photovoltaics and solar thermal energy have the potential however to contribute significant amounts of energy after the year 2000 and development of these technologies should continue.

Reforestation.   In addition to their relationship to the greenhouse effect and climate, forests provide many benefits to life. They provide refuge to most of the planet's living species; they conserve water and soil; they provide a renewable source of fuel and building materials; and they are a source of aesthetic value to humanity. Studies have shown that in some locations there is significantly higher economic value for products from standing forests than from agriculture or other uses of the land. A global effort at reforestation would have many positive effects, in addition to slowing down the rate of CO_2 increase.

Eliminate CFCs. CFCs do not occur in nature; all of them are produced by human activity. The evidence that chlorine derived from CFCs is destroying ozone in the stratosphere is overwhelming; a global ban of all CFC production and use would eventually (over many decades) permit a restoration of the depleted ozone layer.

Plan for Changes and Surprises. Even with all-out efforts of all nations to take action to reduce population growth, conserve energy, reverse the trend of deforestation, and eliminate CFCs, significant environmental change during the next 50 years is still a strong possibility because of human actions of the past. Elimination of the CFCs already in the atmosphere, for example, will take at least a century. Even with an immediate reduction in total CO_2 emissions by as much as 25 percent, the atmospheric concentration of CO_2 would continue to go up because the present rate of removal of CO_2 is about half of the present rate of emissions. Hence it is prudent to plan for possible environmental change such as a further decrease in the amount of ozone, rising sea level, new agriculture practices, changing water supplies and energy requirements, and changing global political and economic threats and opportunities.

It is also prudent to plan for surprises. Although the general threat of CFCs to the ozone layer was predicted in the early 1970s, the ozone hole that was discovered in 1985 came as a surprise. Unpredictable events such as major volcanic eruptions or war could have major effects on climate and the environment. Changes in climate and weather patterns are likely to occur abruptly rather than gradually, and affect the various regions of the Earth in vastly different ways, many of them unpredictable with current scientific understanding and models.

The Cost—Is It Possible? In 1989, Worldwatch Institute estimated the additional cost of slowing population growth, protecting topsoil on farms, reforesting the Earth, and reducing the debt of developing countries. They estimated that these targets could be achieved by *annual* expenditures approaching $50 billion in 1990 increasing to $150 billion by 2000. For comparison, the United Nations Environment Program's Environmental Fund spends approximately $30 million per year while the world spends $1 *trillion* a year on military security, more than $2.7 billion per day!

8.5 SUMMARY

In the last 100 years, humans have increased in number and power to the point at which they are causing significant changes to the atmosphere and to the characteristics of the land surface. Many scientists think that these changes, if unchecked, are likely to cause changes in the weather and climate at a rate unprecedented in human history. Some of the changes may be beneficial locally; however, a global change of the magnitude and rapidity predicted by some would cause additional major stresses to a planet already besieged by increasing human population and activities. Although many of the measures that could be taken to slow down the human impacts on the environment are costly and politically difficult to achieve, they are possible and many of them would prove beneficial even if climate were not to change significantly.

KEY TERMS

albedo	floods	North Pole
albedo effect	fossil fuel consumption	ozone
Antarctica	frequency of extreme events	ozone hole
CO_2 fertilization	glaciers	photodissociation
carbon dioxide	greenhouse gases	polar ice
chlorine	global change	polar vortex
circumpolar circulation	global climate change	regional climate change
cloud forcing	global warming	sea-level rise
computer models	heat waves	skin cancer
Cretaceous period	human population	South Pole
deforestation	ice-albedo feedback	stratosphere
desertification	melanoma	water vapor-temperature
drought	methane	feedback
feedbacks	Montreal protocol	

PROBLEMS

1. Give two examples each of positive and negative feedbacks in the climate system.

2. The enhanced greenhouse effect by itself would certainly lead to a warmer climate globally. Explain why any negative feedback initiated by the enhanced greenhouse effect can only reduce the warming, not eliminate it altogether or produce a global cooling.

3. If the polar regions warm up more than the tropical regions due to the enhanced greenhouse effect, would you expect the polar jet stream to weaken or intensify, on the average?

4. If the oceans and lower troposphere in the tropics become warmer, would you expect hurricanes to become stronger or weaker? Why?

5. Name two good environmental reasons for eliminating CFCs altogether.

6. Explain the regular oscillations superimposed on the overall upward trend of carbon dioxide shown in Figure 8.4.

7. Why is reforestation not a permanent solution to the problem of increasing carbon dioxide in the atmosphere?

8. Name two processes that could cause a global rise of sea level.

9. Why did the "ozone hole" first form over Antarctica? Why is a similar hole much less likely to form over the continental United States?

10. Surface temperatures in many parts of the world vary by more than 20°C from summer to winter and life has adapted well to these variations. Why then are people concerned about the possibility of a mean global surface temperature increase of 2–3°C over the next 50 years?

11. Explain how an increase in global mean cloud cover might result in either a mean global surface warming or cooling.

Appendix I:
Units Used in This Book

TABLE I.1   Temperature Scales

	Fahrenheit (°F)	Celsius (°C)	Kelvin (K) or Absolute (A)
Boiling point of water	212	100	373
Melting point of ice	32	0	273
Divisions between fixed points	180	100	100
Conversion formulas:	$\dfrac{°F - 32}{180} = \dfrac{°C}{100}$; $K = °C + 273$		

TABLE I.2   Length

1 kilometer = 0.6214 statute mile = 0.5396 nautical mile
1 meter (m) = 1.093611 yards = 3.2808 feet = 39.370 inches
1 cm = 0.3937 in = 10^4 micrometers μm) = 10^8 angstroms (Å)
1 mile = 1.61 km
1 micrometer (μm) = 10^{-6}m

TABLE I.3   Velocity

1 knot (nautical mile per hour) = 1.1516 statute mi/h = 0.5148 m/s = 1.857 km/h
1 mi/h = 0.8684 knot = 0.447 m/s
1 m/s = 2.2369 mi/h = 1.9424 knots = 3.2808 ft/s
1 km/h = 0.62 mi/h

TABLE I.4   Force

1 newton (N) = 1 kg–m/s^2
1 dyne = 10^{-5}N
1 pound = 4.4482 N (British)

TABLE I.5   Pressure

1 pascal (Pa) = 1 newton/m^2 (N/m^2) = 10^{-2} millibar (mb)
1 mb = 0.02953 inches of mercury (in Hg) = 1000 dynes/cm^2
1 in Hg = 33.8639 mb
1 lb/in^2 = 68.947 mb

TABLE I.6   Energy

1 gram-calorie [or, just "calorie," (cal)] --- amount of energy required to raise the temperature of 1 g of water 1°C.

1 erg = 1 dyne cm = 2.388×10^{-8} cal

1 watt-hour = 860 gram-calories (g-cal) = 3.600×10^{10} ergs

1 British thermal unit (Btu) = 0.293 watt-hour = 251.98 gram-calories = 1.055×10^{10} ergs

1 joule (J) = 10^7 ergs

1 cal = 4.1855×10^7 ergs

1 foot-pound = 1.356×10^7 ergs

1 horsepower-hour = 2.684×10^{13} ergs = 0.6416×10^6 cal

1 kilowatt-hour = 3.6×10^6 J

TABLE I.7   Power

1 watt = 14.3353 cal min.$^{-1}$

1 cal min.$^{-1}$ = 0.06976 watt

1 horsepower = 746 watts

1 Btu min.$^{-1}$ = 175.84 watts = 252.08 cal min.$^{-1}$

TABLE I.8   Powers of Ten

10^{-2} = 0.01

10^{-1} = 0.1

10^0 = 1.0

10^1 = 10

10^2 = 100

TABLE I.9   Prefixes

Factors by Which Unit is Multiplied	Prefix	Symbol
10^{12}	tera	T
10^9	giga	G
10^6	mega	M
10^3	kilo	k
10^2	hecto	h
10	deka	da
10^{-1}	deci	d
10^{-2}	centi	c
10^{-3}	milli	m
10^{-6}	micro	μ
10^{-9}	nano	n
10^{-12}	pico	p
10^{-15}	femto	f

Appendix II:
Standard Atmosphere

Altitude (m)	Temperature (°C)	Pressure (mb)	Density (kg/m³)	Altitude (m)	Temperature (°C)	Pressure (mb)	Density (kg/m³)
0	15.0	1,013.2	1.2250	11,000	−56.4	227.0	0.3648
500	11.8	954.6	1.1673	11,100	−56.5	223.5	0.3593
1,000	8.5	898.8	1.1117	11,500	−56.5	209.8	0.3374
1,500	5.2	845.6	1.0581	12,000	−56.5	194.0	0.3119
2,000	2.0	795.0	1.0066	13,000	−56.5	165.8	0.2666
2,500	−1.2	746.9	0.9569	14,000	−56.5	141.7	0.2279
3,000	−4.5	701.2	0.9092	15,000	−56.5	121.1	0.1948
3,500	−7.7	657.8	0.8634	16,000	−56.5	103.5	0.1665
4,000	−11.0	616.6	0.8194	17,000	−56.5	88.5	0.1423
4,500	−14.2	577.5	0.7770	18,000	−56.5	75.6	0.1216
5,000	−17.5	540.5	0.7364	19,000	−56.5	64.7	0.1040
5,500	−20.7	505.4	0.6975	20,000	−56.5	55.3	0.0889
6,000	−24.0	472.2	0.6601	25,000	−51.6	25.5	0.0401
6,500	−27.2	440.8	0.6243	30,000	−46.6	12.0	0.0184
7,000	−30.4	411.0	0.5900	35,000	−30.6	5.7	0.0085
7,500	−33.7	383.0	0.5572	40,000	−22.8	2.9	0.0040
8,000	−36.9	356.5	0.5258	45,000	−9.0	1.5	0.0020
8,500	−40.2	331.5	0.4958	50,000	−2.5	0.8	0.0010
9,000	−43.4	308.0	0.4671	60,000	−17.4	0.225	0.000306
9,500	−46.7	285.8	0.4397	70,000	−53.4	0.055	0.000088
10,000	−49.9	265.0	0.4140	80,000	−92.5	0.010	0.000020
10,500	−53.1	245.4	0.3886	90,000	−92.5	0.002	0.000003

Appendix III: Plotting Model for Sea-Level Weather Chart

TABLE III.1   WW Symbols

Weather type	Symbol
haze	∞
fog	☰
drizzle	,
rain	•
continuous light rain	✱
snow	▽̇
rain shower	⍑
thunderstorm	⍑̇
snow shower	▽̇

Symbols		Example
N:	Amount of total sky cover	Overcast
ff:	Barbs show wind speed (full barb = 10 knots)	25 knots
dd:	Arrow shaft shows wind direction	Northwest
TT:	Temperature (°F)	38°F
VV:	Visibility (miles)	1.5 miles
WW:	Weather type	Continuous light rain
T_dT_d:	Dew point temperature (°F)	34°F
C_l:	Type of low clouds	stratus
h:	Height of ceiling	300–599 ft
N_h:	Amount of low cloud cover	6/8
RR:	Precipitation amount, past 6 hours	0.26 in.
W:	Weather, past 6 hours	Rain
R_t:	Time precipitation began or ended	Began 3–4 hours ago
a:	Trend of barograph curve, past 3 hours	Rising
pp:	Pressure change, past 3 hours in mb multiplied by 10	+ 2.2 mb
PPP:	Sea level pressure, with only last three digits (including tenths) given	1,013.2 mb
C_m:	Type of middle clouds	Nimbostratus
C_h:	Type of high clouds	Cirrus

Appendix IV: Plotting Model for Upper-Air Chart

	Symbols	Example
dd:	Arrow shaft shows wind direction	270°
ff:	Barbs show wind speed (triangle, 50 knots; full barb, 10 knots)	65 knots
TT:	Temperature (°C)	–29°C
T_dT_d:	Temperature—dew-point difference (°C)	10°C
hhh:	Height of pressure surface (in meters), with only first three digits given	5400 m (500-mb surface)

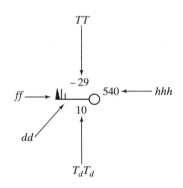

Appendix V:
Supplementary Readings

BOOKS ON GENERAL METEOROLOGY AND CLIMATOLOGY

Atkinson, B. W., ed. *Dynamical Meteorology, An Introductory Selection.* Agincourt, Canada: Methuen Publications, 1981.

————. *Mesoscale Atmospheric Circulations,* New York: Academic Press, 1981.

Bentley, W. A., and **W. J. Humphreys.** *Snow Crystals.* New York: Dover Publications, 1962.

Calder, Nigel. *The Weather Machine.* New York: Viking Press, 1974.

Cotton, W., and **R. Anthes.** *Storm and Cloud Dynamics.* New York: Academic Press, 1989.

Firor, J. *The Changing Atmosphere: A Global Challenge,* New Haven: Yale University Press, 1990.

Fishman, J., and **R. Kalish** *Global Alert: The Ozone Pollution Crisis.* New York: Plenum Press, 1990.

Fleagle, R. *Global Environmental Change. Interactions of Science, Police, and Politics in the United States,* Westport Connecticut: Praeger Press, 1994.

Frisinger, H. H. *The History of Meteorology: to 1800.* Boston: American Meteorological Society, 1977.

Greenler, R. *Rainbows, Halos and Glories.* New York: Cambridge University Press, 1989.

Griffiths, S. F., and **D. M. Driscoll.** *Survey of Climatology.* Columbus: Charles E. Merrill Publishing Co., 1982.

Hare, F. K., and **M. K. Thomas.** *Climate Canada.* 2d ed. New York: John Wiley and Sons, 1979.

Hughes, P. *American Weather Stories.* Washington D.C.: U.S. Department of Commerce, 1976.

Kirk, R. *Snow.* New York: William Morrow and Company, 1978.

Kutzbach, G. *The Thermal History of Cyclones: A History of Meteorological Thought in the Nineteenth Century.* Boston: American Meteorological Society, 1979.

Lockhart, G. *The Weather Companion: An Album of Meteorological History, Science, Legend, and Folklore.* New York: John Wiley and sons, 1988.

Ludlum, D. M. *Weather Record Book, United States and Canada.* Washington, D.C.: Weatherwise Inc., 1971.

————. *Early American Hurricanes 1492–1870.* Boston: American Meteorological Society, 1963.

————. *Early American Winters 1604–1820.* Boston: American Meteorological Society, 1966.

————. *Early American Winters II 1821–1870.* Boston: American Meteorological Society, 1968.

————. *Early American Tornadoes 1586–1870.* Boston: American Meteorological Society, 1970.

Lutgens, F. K., and **E. J. Tarbuck.** *The Atmosphere*. 6th ed. Englewood Cliffs, N.J.: Prentice-Hall, Inc. 1989.

Lynch, D. K., ed. *Atmospheric Phenomena: Readings from Scientific American*. San Francisco: W. H. Freeman and Co., 1979.

Maunder, W. J. *The Value of Weather*. Agincourt, Canada: Methuen Publications, 1970.

Moran, J. M., and **M. D. Moran.** *Meteorology: The Atmosphere and the Science of Weather*. 4th ed. New York: Macmillan, 1994.

Pielke, R. *The Hurricane*. New York: Routledge, 1990.

Ruffner, J. A., and **F. E. Bair.** *The Weather Almanac*. 2d ed. New York: Avon Publications, 1977.

Schaefer, V. J., and **J. A. Day** *A Field Guide to the Atmosphere*. Boston: Houghton Mifflin Co., 1981.

Scorer, R. *Clouds of the World: A Complete Color Encyclopedia*. North Pomfret, Vt.: David and Charles, 1972.

Simpson, R. H., and **H. Riehl** *The Hurricane and Its Impact*. Baton Rouge, La: Louisiana State University Press, 1981.

Trewartha, G. T., and **L. Horn.** *An Introduction to Climate*. 5th ed. New York: McGraw Hill Book Co., 1979.

Unman, M. A. *Understanding Lightning*: Pittsburgh: Bek Industries, 1971.

United States Committee for Global Atmospheric Research Program. *Understanding Climatic Change, A Program for Action*. Washington, D.C.: National Academy of Sciences, 1978.

Winkless, Nels, III, and **I. Browning.** *Climate and the Affairs of Men*. New York: Harper's Magazine Press, 1975.

ADDITIONAL SOURCES OF WEATHER DATA AND USEFUL PERIODICALS

The Smithsonian Institution, Washington, D.C., World Weather Records.

National Climatic Center (NOAA), Asheville, N.C.
    Average Monthly Weather Resume and Outlook
    Climates of the States
    Climatic Charts for the United States
    Climatological Data for the U.S. by Sections
    Daily Weather Map
    Monthly Climatic Data for the World
    Storm Data

Weather, published monthly by the Royal Meteorological Society, London, England.

Weatherwise, published bimonthly by Heldref Publications, Washington, D.C.

Bulletin of the American Meteorological Society, published monthly by the American Meteorological Society, Boston, MA.

Index